Nuclear Materials

Edited by Pavel V. Tsvetkov

Published in London, United Kingdom

IntechOpen

Supporting open minds since 2005

Nuclear Materials
http://dx.doi.org/10.5772/intechopen.83315
Edited by Pavel V. Tsvetkov

Contributors
Douglas Burns, Stephen Johnson, Viacheslav Sergeevich Okunev, Analía Soldati, Miguel Oscar Prado, Diana Carolina Lago, Pingzhou Ming, Nikolay A. Andreevich Makhutov, Igor A. Razumovsky, Sergey V. Maslov, Mikhail M. Gadenin, Dmitry O. Reznikov, Edward Jimenez, Pavel V. Tsvetkov

First published in London, United Kingdom, 2021 by IntechOpen
IntechOpen is the global imprint of INTECHOPEN LIMITED, registered in England and Wales, registration number: 11086078, 5 Princes Gate Court, London, SW7 2QJ, United Kingdom
Printed in Croatia

British Library Cataloguing-in-Publication Data
A catalogue record for this book is available from the British Library

Additional hard and PDF copies can be obtained from orders@intechopen.com

Nuclear Materials
Edited by Pavel V. Tsvetkov
p. cm.
Print ISBN 978-1-83962-371-4
Online ISBN 978-1-83962-372-1
eBook (PDF) ISBN 978-1-83962-373-8

For EU product safety concerns:
IN TECH d.o.o., Prolaz Marije Krucifikse Kozulić 3, 51000 Rijeka, Croatia, info@intechopen.com or visit our website at intechopen.com.

Meet the editor

Pavel V. Tsvetkov, Ph.D., is an Associate Professor in the Department of Nuclear Engineering, Texas A&M University. Dr. Tsvetkov's research program is focused on novel energy systems meeting global growing needs in sustainable resources. His project portfolio includes direct energy conversion, waste-minimization efforts, novel reactor designs, instrumentation efforts, and data science and engineering for a broad range of applications targeting optimized designs and performance. Dr. Tsvetkov is a member of the American Nuclear Society (ANS), American Society of Mechanical Engineers (ASME), American Society of Engineering Education (ASEE), Alpha Nu Sigma and Phi Kappa Phi. He has published more than 300 papers in peer journals, conference proceedings, and reports as well as served as an editor and major contributor for fourteen books on energy, environment, and nuclear energy.

Contents

Preface

Advancements in nuclear technology are inherently linked to advancements in related materials development, manufacturing, and availability. This book offers readers the opportunity to explore contemporary and emerging frontiers in nuclear materials, showcasing technologies and presenting reasons behind engineering solutions.

This book is not meant to be a comprehensive resource on the subject but a collection of select contributions on topics of interest within the nuclear engineering community. The spectrum of covered topics ranges from fundamentals to such applied topics as accident-tolerant nuclear fuels and materials performance in various radiation environments to nuclear materials management.

Articles on materials compatibility, corrosion, and radiation damage survey developments of paramount importance for current and novel nuclear systems and applications. Manufacturing options and performance evaluation metrics and methods are included.

Readers are expected to gain insights into the status of nuclear materials, underlying challenges, paradigm-shifting breakthroughs, and trailblazers. The discussion of nuclear materials goes beyond nuclear energy to everyday nuclear material uses.

Let's look into the art of nuclear materials, fundamentals, and applications to create and design new nuclear energy systems and enhance contemporary technologies!

Pavel V. Tsvetkov
Department of Nuclear Engineering,
Texas A&M University,
College Station, Texas, USA

Section 1

Introduction

Chapter 1

Introductory Chapter: The Key Role of Materials in Nuclear Technology Options and Pathways

Pavel V. Tsvetkov

1. Introduction

In this book we are focusing on a broad range of nuclear technologies. In particular, we would like to give special attention to energy systems. The feasible way to utilize nuclear energy today is via deployment of fission reactors. While fusion reactors are technically possible and their physics has been demonstrated, it will take a much longer time to bring fusion systems into the feasible and competitive energy engineering domain of commercially viable systems. However, material developments targeting performance under extreme conditions is relevant to both pathways enabling options for fission and fusion technologies. As such, this book discusses developments broadly relevant to nuclear energy systems, fission and fusion.

Materials play the key role in the nuclear energy system feasibility. So much so, they are among the key classification factors and metrics for nuclear systems:

Nuclear system type (fission or fusion)

- Purpose and functionality,
- Moderator type,
- Coolant type,
- Fuel type,
- Structural materials options,
- Neutron-energy classification,
- Core design options (homogeneous or heterogeneous, and etc.),
- Energy conversion process type and implementation options,
- Environmental interfaces.

Focusing on nuclear fission systems as the near term and already available commercial option, there are a few well-explored directions as well as emerging new technologies. Materials play the enabling role for contemporary nuclear reactors creating possibilities for extending their lifetimes. Materials are also the key factor making emerging new technologies viable. As notable examples, there are developments of paramount significance offering accident tolerant fuels and robust structural materials performing under extreme conditions due to high temperatures and high energy radiation effects.

2. Performance conditions and materials options

There are many options for material selections. The contemporary nuclear reactors utilize mostly light water (light water reactors) that is prevented from boiling (in pressurized water reactors) or allowed to boil (in boiling water reactors). All nuclear reactors utilizing water pressurize their primary system to achieve needed performance characteristics. The pressure levels range from about 6 MPa in boiling water reactors to 15 MPa in pressurized water reactors. A number of contemporary commercial nuclear reactors utilize heavy water (CANDU reactors).

The alternative options for primary coolant choices are liquid metals, such as molten sodium, and gasses, such as helium. Respectively, the reactor types are called liquid-metal-cooled reactors and gas-cooled reactors. Depending on the choice of a liquid metal, an expensive intermediate loop might be needed to isolate the working-fluid energy conversion loop from high radioactivity levels induced in the primary loop. The dramatically different option is offered by liquid salt configurations found in molten salt reactors. In all of these advanced reactor options, material interactions, compatibility, and performance in their ability to support and withstand internal conditions while offering needed characteristics are vital for successful developments towards increasing commercial feasibility levels in competition with contemporary options.

Moderators and structural materials include both solid and liquid options. Graphite, beryllium, steels, and composites are among solid form options. Light and heavy water and liquid salts are among solid form options. The choice of moderators and structural materials is driven by internal conditions and performance characteristics. Most of the time, materials are expected to remain compatible and perform under extreme conditions over prolonged periods of time.

Some material choices offer unique traits and opportunities not found in others. For example, due to characteristics of liquid metals such as high boiling points and very low vapor pressures, the pressurization is not required or very low. This is an important distinction and advantage compare to light water and gasses which do require pressurization to achieve needed performance characteristics in a system. As noted above, typical pressures in light water systems are between 6 and 15 MPa. Typical pressures in gas systems are between 4 and 7 MPa. The need to support these pressure levels poses requirements on structural materials for primary systems including vessels for components and connecting piping. Absence of the need for pressurization is a significant relaxation of the demand on materials.

Notably, liquid salts take advanced nuclear systems even further by eliminating requirements for solid structural materials to withstand prolonged direct proximity to nuclear fuel. This offers significant advantages from the system longevity point as well as from the system safety point. Molten salt reactors do require unique technologies in support of salt environments that would not be found in other nuclear reactor configurations.

3. Materials selection and the system design

The choice of materials in nuclear energy systems determines attainable neutron energies within their reactor cores. Consequently, it defines these systems as either thermal or fast nuclear reactors or systems. The contemporary nuclear fleet is predominantly composed of thermal reactors although a number of fast reactors have been built and operated over the years, several fast reactors are in operation today. Thermal reactors use water, heavy water, helium, carbon dioxide, and graphite as material options. Fast reactors use sodium, lead, and steels.

Considering the importance of fast reactors for sustainable nuclear energy future pathways, let us summarize advantages of liquid metals over other in-core material choices accounting for their heat removal capabilities:

- Liquid metals have excellent heat-transfer characteristics.
- Liquid metals are characterized by wide ranges of temperatures in which they remain in the liquid state and can offer high temperature performance matching characteristics of gas-cooled reactors.
- Liquid metals, helium, and heavy salts are the coolant types for fast reactors.
- Liquid metals have excellent resistance to nuclear radiation damage.
- Liquid metals have high thermal conductivities and low specific heats due to the temperature gradients in the coolant system to be low. Coupled with high boiling temperatures, local hot spots are inherently minimized in fast reactor configurations.

From the challenges side, liquid metals are chemically active and corrosive, requiring the use of special, and often costly, structural materials and handling technologies. Oxygen, present even in small quantities, oxidizes sodium to Na_2O, which is highly soluble in Na. It later precipitates on cold walls and causes clogging problems. The relatively high freezing point of sodium necessitates the use of electric or other heaters to keep the coolant from freezing during low-power operation or extended shutdown. In addition, liquid metals are not universally available and are costly. All of these challenges are engineering challenges. They have been overcome and resolved through the use of advanced materials designed to perform in liquid metal environment.

Gas coolants is another option for advanced nuclear reactors. Because they have very small moderation capabilities at reactor pressures, separate moderators, such as graphite or heavy water, are needed in thermal reactors. Although there have never been a gas-cooled fast reactor in operation, there are significant interests in industry to develop and deploy a number of gas-cooled fast reactor technologies. Gaseous coolants are generally available, cheap, safe, and manageable. They allow operation modes with high reactor outlet temperatures, resulting in high plant thermal efficiencies. Furthermore, gas coolants allow high-efficiency direct thermodynamic cycles. When purified, they do not present a serious activation problem. Gases, however, have poor heat-transfer characteristics and low volumetric heat capacities, and they require greater pumping powers and larger ducts than do liquid coolants. Pressurizing is necessary to reduce pumping requirements. Leak-proof systems are needed, especially for low-molecular-mass gases such as He. Because of the poor heat transfer, high fuel temperatures are required if high heat-removal rates from the reactor are to be achieved. Similar to liquid metal configurations, these challenges are engineering challenges. They have also been resolved by development and deployment of advanced materials that are suitable for gas environments.

Thus, it can be asserted that the very selection of materials and materials availability are the key factors determining nuclear reactor feasibility, both from technology side and from commercial deployment side. Fortunately, advances in materials make a wide range of nuclear reactors possible today. This book describes many key contemporary developments in nuclear materials targeting contemporary and advanced nuclear energy systems as well as outlines the enabling technologies and approaches for the future.

Author details

Pavel V. Tsvetkov
Department of Nuclear Engineering, Texas A&M University, College Station, Texas, USA

*Address all correspondence to: tsvetkov@tamu.edu

Section 2

Advanced Fuel Materials

Chapter 2

Uranium Dioxide Nanoparticulated Materials

Analía Leticia Soldati, Diana Carolina Lago and Miguel Oscar Prado

Abstract

Nanostructured actinide materials have gained the attention of the nuclear community after the discovery of enhanced properties in fuels that undergo high burn up. On these conditions, the UO_2 grains experimented recrystallization and formed a new rim of UO_2 nanoparticles, called high burn up structures (HBS). The pellets with HBS showed closed porosity with better fission gas retention and radiation tolerance, ameliorated mechanical properties, and less detriment of the thermal conductivity upon use. In this chapter, we will review different ways to obtain uranium nanoparticles, with emphasis on their synthesis and characterization. On the one hand, we will comment on radiation chemical syntheses, organic precursor-assisted syntheses, denitration processes, and biologically mediated syntheses. On the other hand, we will include for each of them a reference to the appropriate tools of the materials science that are used to fully characterize physical and chemical properties of these actinide nanoparticles.

Keywords: UO_2, nanoparticles, grain sizes, synthesis, characterization

1. Introduction

Nanomaterials, which are present naturally in the environment and also as a result of anthropogenic activities (incidental or engineered), gain the attention of scientist and technologists due to their promising applications. The surface-to-volume ratio, grain size, morphology, composition and elemental distribution affect nanoparticle's physicochemical and electrical properties, surface reactivity, material growth, or dissolution rates [1]. These characteristics can be thus engineered to take advantage of the nanoparticles over their macroscopic equivalents, for example, to favor faster catalysis of reactions, high loading of medicines or absorption of toxins from polluted zones.

In the nuclear material's field, actinide oxides nanoparticles became under systematic study after the detection of two main issues:

First, the discovery of a rim structure in UO_2 pellets that had have a burn up of 40–67 GWd/tM (also called high burn up structures or HBS [2]). The pellet, initially formed by micrometer-sized grains recrystallized in a ring of nanoparticles at the rim. The pellets with HBS presented better fission gas retention, ameliorated radiation tolerance and mechanical properties as the plasticity [3]. The direct consequence of this observation was an increment in the number of publications dealing with different synthesis of UO_2 nanoparticles to form pellets mimicking from the beginning the HBS structure [3–7].

IntechOpen

Second, the fact that actinides tend to form colloids of aggregated nanoparticles [8, 9]. Indeed, in contact with water, metallic U corrosion is known to form fine UO_2 particulates [10, 11]. This material has different properties than micrometer particulated material, affecting, for example, the expected behavior in spent nuclear fuels, radioactive wastes, and contaminated places, due to their differences in mobility, solubility, surface reactivity, complexation, speciation, weathering, eco-toxicity, and biological uptake. In particular, because their small size, nanoparticles may have a toxic effect on living organisms that is not present with micrometer-sized particles. Thus, there is a need for expanding the actual knowledge on actinide nanoparticles with emphasis in their physicochemical properties, grain sizes, crystal phases, elemental distribution and reactivity, for predicting and controlling their behavior under different conditions. This knowledge will also serve to redesigning long-term nuclear waste disposals and mobility barriers.

Both former topics request well-characterized actinide nanoparticles, especially those composed of UO_2. That, added to the scientific motivation per se, is represented in the increased number of publications in the past 25 years dealing with different synthesis and characterization of UO_2 nanoparticles. In the next sections we resume and discuss different methods to obtain particles of uranium dioxide with grain sizes in the sub-micrometer range. We divided the methods by the type of synthesis. On one side, there are those which follow a wet chemical route, subdivided in processes that use a wet denitration step and processes which need an organic precursor, such as variation of sol-gel or Pechini syntheses. On the other side, we explain those methods which use irradiation with particles or photons to induce UO_2 particle formation. In addition, we describe biologically assisted syntheses, which make use of cells and bacteria to precipitate UO_2 nanoparticles.

It is worth to mention at this point that many of the published syntheses in articles or patents were focused to the production of UO_2 for its use in nuclear reactors. This application requires a powder with good fluency and compressibility to further handling for pellet fabrication. Thus, fractions of particles with sub-micrometer diameter, which sometimes are referred as "very fine powder," were separated from the bulk and discarded. In addition, very often nanoparticles aggregate in micrometer-sized particles. Only with high-resolution microscopy techniques, or indirectly through BET surface area measurements, for example, it is possible to detect the nanometric structure of the material. Therefore, in more than one publication, nanoparticles are wrongly classified as micrometer-sized particles. Here we attract the attention on this fact in some of the reported works.

2. Chemical and electrochemical routes

2.1 Syntheses from inorganic uranyl salts

In the group of the wet chemical syntheses, one of the most common practices to obtain UO_2 to manufacture nuclear fuel pellets is the physicochemical precipitation, followed by calcination and reduction [12]. The ammonium di-uranate (ADU) and the ammonium uranyl carbonate (AUC) routes are two well-known examples. Both start from an inorganic uranium salt such as the uranyl nitrate hexahydrate (UNH), involve thermal treatments in different atmospheres and, at intermediate to high temperatures, obtain the fluorite fcc UO_2 phase.

Although the ADU synthesis originally was not tuned to produce nanoparticles, first studies describe that depending on pH and synthesis conditions, a fine powder with sub-micrometer structure and a grain size of 370 nm was observed [13]. Some years ago, Soldati et al. took advantage of characterization methods from the nanoscience and demonstrated that the UO_2 particles obtained by the ADU route in the standard conditions described elsewhere (i.e., pH 9 and 60°C thermal bath) were indeed agglomerates of rounded, but irregular, nanoparticles of homogeneous composition, fcc Fm-3 m crystal phase, and 80–120 nm crystal sizes [4]. In that experience, to obtain about 100 g UO_2 nanoparticles with those characteristics by the ADU method requires a filtrating step, produces about 2 L ammonia water waste, needs 12–16 hours thermal treatments at intermediate to high temperatures, and consumes air and a reducing atmosphere such as H_2:Ar (10:90) [4, 13].

In these syntheses, UO_2, and some mixed oxides with Gd or Pu, can be obtained from a solution of the actinides (as nitrates or oxides) in 1 M HNO_3, concentrations of 50–400 g/L, 60°C, and pH between 4 and 9 [4, 13–16]. The precipitation of ADU is favored by mixing the mother solution with a basic 13 M (NH_4OH) solution [14, 15, 17] or bubbling NH_3 gas [4, 13] (Eqs. (1)–(3)).

For example, for the case of ADU, the involved reactions are:

$$NH_3{}^{g} + H_2O \rightarrow NH_4^{+} + OH^{-} \tag{1}$$

$$2UO_2^{+2} + 6OH^{-} \rightarrow U_2O_7^{-2} + 3H_2O \tag{2}$$

$$U_2O_7^{-2} + 2NH_4^{+} \rightarrow U_2O_7(NH_4)_2 \downarrow_{(ADU)} \tag{3}$$

Once that the precipitated phase is completely formed, the solution is stirred for 1 hour and vacuum filtrated, washed with milliQ water, and dried between 80 and 120°C for 24 hours. After that, the ADU is converted to U_3O_8 by calcination at 800°C in air for 6–8 hours (Eqs. (4) and (5)).

$$U_2O_7(NH_4)_2 \xrightarrow{400°C,\ air} 2UO_3 + 2NH_3\uparrow + H_2O\uparrow \tag{4}$$

$$3UO_3 \xrightarrow{400°C\ to\ 800°C,\ air} U_3O_8 + \tfrac{1}{2}O_2 \tag{5}$$

Finally, the U_3O_8 is reduced to UO_2 by thermal treatment between 650 and 700°C for 7 hours in pure H_2 or mixtures of H_2 and Ar or N_2 in proportions of 8–10% (Eq. (6)).

$$U_3O_8 + 2H_2 \xrightarrow{650°C,\quad (10:90)\ H_2:Ar} 3UO_2 + 2H_2O\uparrow \tag{6}$$

On the other side, the AUC, for example, is precipitated from the UNH-HNO_3 solution with $(NH_4)_2CO_3$ [14] and converted to UO_2 at 650°C in a water vapor/hydrogen atmosphere. However, to the best of our knowledge, only micrometric particle sizes were reported by AUC syntheses.

2.2 Syntheses from organic uranyl salts

An alternative way for precipitating UO_2 nanoparticles from the inorganic salt uranyl nitrate are the synthesis from the organic salts uranyl acetylacetonate (UAA) or acetate (UA), mediated by organic solvents and temperature. Wu et al., for example, obtained 3–8-nm-large cubic UO_2 nanocrystals by decomposition at 295°C, under Ar, of UAA in a mixture of oleic acid (OA), oleylamine (OAm), and octadecene (ODE) [18]. Non-agglomerated and highly crystalline UO_2 particles were obtained in a similar synthesis by Hudry et al. at temperatures of 280°C [19]. These nanoparticles were isotropic faceted nanodots of 3.6 ± 0.4 nm diameter. Moreover, Hu et al. used UA dissolved in oleylamine (OAm) and oleic acid (OA) which after heating in an oil bath, centrifuging, washing with ethanol, and dispersing in cyclohexane resulted in two-dimensional nanoribbons of U_3O_8 with dimensions of about 4 × 100 nm. Higher autogenous pressure, in an autoclave, was useful for obtaining wider nanoribbons. With the addition of octadecene (ODE) or toluene, U_3O_7 nanowires were obtained whose width is about 1 nm and length varied in the range of 50–500 nm depending on the temperature-time conditions of the process [20]. In addition, sphere-shaped UO_2 nanoparticles with an average diameter of 100 nm, which consisted in 15 nm nanocrystal subunits, were obtained by Wang et al. from a 0.5 mM UA aqueous solution mixed with ethylenediamine, autoclaved, and heated at 160°C for 48 h [21]. On the other hand, Tyrpekl et al. obtained 5–11 nm UO_2 nanoparticles by annealing a dry precipitate of $(N_2H_5)_2U_2(C_2O_4)_5 \times nH_2O$ at 600°C in Ar [22].

2.3 Sol-gel syntheses

A colloid is a suspension in which the dispersed phase particle's size is so small (~1–1000 nm) that gravitational forces are negligible and interactions are dominated by short-range forces, such as van der Waals attraction and surface charges. In the context of the sol-gel synthesis, the "sol" is formed by a colloidal suspension of solid particles in a liquid, while the "gel" is a suspension of a liquid phase in a continuous solid phase [23]. Basically, two sol-gel routes are used: the polymeric route using alkoxides and the colloidal route using metal salts.

In a typical polymeric sol-gel process, as the one used for low-temperature preparation of SiO_2 monoliths from a tetraethoxisilane (TEOS) solution, a polymerized structure is formed by the condensation of alcohols proceeding from the TEOS hydrolysis. Another widely used sol-gel synthesis is the complexation by amines, known as internal gelation [24–27]. This synthesis is common to find in the nuclear field associated to the fabrication of UO_2 microspheres formed by agglomerated nanoparticles as in the work of Daniels et al. [25]. In this case, an uranyl nitrate solution is mixed with urea ($CO(NH_2)_2$) and hexamethylenetetramine (HMTA) solution. Then, the HMTA is decomposed at low temperature (90°C) causing an increase in pH and hydrolysis of uranium (Eqs. (7) and (8)), resulting in a solution gelation:

$$\text{Hydrolysis}: UO_2^{+2} + 2H_2O \leftrightarrow UO_2(OH)_2 + 2H^+ \quad (7)$$

$$\text{Condensation}: UO_2(OH)_2 + H_2O \rightarrow UO_3 \bullet 2H_2O \downarrow \quad (8)$$

This gel is washed with NH_4OH and dried to obtain dry UO_3. Later, thermal treatments at 800°C in air allow obtaining U_3O_8 powders that are further reduced to UO_2 particles. With this method, UO_2 millimeter-sized spheres with a nanometric substructure were obtained by different authors [25, 26]. The powder morphologies and particle sizes depend on the temperature and the calcination atmospheres used. The average particle size varies between 100 and 4000 nm. The samples obtained through the oxalic route and a single calcination (in neutral or reductive atmosphere) showed similar lattice parameters, close to the value of UO_2 [24].

Recently Leblanc et al. presented another method that they called "advanced thermal denitration in presence of organic additives" that includes a gelation step of the uranyl nitrate solution [28]. In this process, an acidic uranyl nitrate solution is prepared, and urea is added to avoid uranium precipitation. Oxide synthesis was performed by adding two monomer types: acrylic acid (AA) and N,N′-methylene bis acrylamide (MBAM) in a molar ratio of 20:1 (AA:MBAM). A fully homogeneous solution was obtained, which when heated up to 100°C and after the addition of 25 mL of hydrogen peroxide (30 wt%) as initiator completely polymerized into a gel. The entire solution is incorporated into the polymer network, ensuring that all the cations of the system are stripped into the obtained gel. Drying at 150°C, following an oxidative calcination of the organic part at 800°C, and finally reducing it in Ar:5%H_2 at 800°C resulted in a nanostructured material with crystallite size below 100 nm, as determined by XRD diffraction.

2.4 Electrochemically assisted syntheses

Recently, Rousseau et al. presented a wet chemical novel method to synthesize UO_2 (and also UO_2 doped with tetra- or trivalent elements), based on the electrochemical reduction of U^{6+} to U^{4+}, followed by a precipitation in a reducing and anoxic condition, at constant pH [29]. The mother U^{4+} solution was made dissolving UNH in 1 M NaCl. The authors studied two methods for precipitating stable UO_{2+x} nanoparticles of different sizes. In the pH range 2.5–4, the starting U^{6+} solution was added to the NaCl solution under reducing conditions, and U^{6+} cations were reduced electrochemically to U^{4+}. The increment in pH was compensated with 0.1 M HCl. In the pH range 4–8, the mother U^{6+} solution was added drop by drop directly to the 1 M NaCl solution, balancing the pH change with 0.1 M NaOH. A redox potential of −300 mV/NHE was applied using Pt electrodes. The obtained products were filtered with a 0.22 μm filter, and the precipitates were washed two times with ultrapure water. The nanoparticles produced correspond to a single fluorite $UO_{2.19\,\pm\,0.01}$ phase and average TEM coherent domain size of (12 ± 2) nm for pH < 4 and $UO_{2.11\,\pm\,0.02}$ of 4–6 nm for pH 6.5. The BET surface area for this nanomaterial was 10.3 ± 0.1 m^2/g, which the authors associated to a grain size of 53 nm, indicating a moderate agglomeration of the nanoparticles. XPS, in good agreement with the other analytical techniques, resulted in a U^{6+}/U^{4+} ratio close to 0.1.

Moreover, an electrolytically reduced aqueous solution of 0.5 M uranyl nitrate was used as precursor, together with NaOH solution as alkalinization agent, to trigger the precipitation of UO_2 nanoparticles near the U^{4+} solubility line. XRD and HR-TEM analyses showed that the nanoparticles obtained exhibit the typical slightly oxidized UO_{2+x} fcc fluorite structure, with an average crystal size of 3.9 nm and a narrow size distribution [6].

In these cases, the reduction is mediated by the reactions occurring in the cathode (Eq. (9)) and in the anode (Eq. (10)), respectively [6]:

$$UO_2^{2+} + 4H^+ + 2e^- \leftrightarrow U^{4+} + 2H_2O \quad (9)$$

$$2H_2O \leftrightarrow 4H^+ + 4e^- + O_2 \quad (10)$$

To maintain the reducing environment, the oxygen must be eliminated with an oxygen-free gas such as pure Ar. In the work of Jovani-Abril et al. [6], for example, the starting pH was 0.5, and the solution was slowly alkalinized to allow the precipitation of the UO_2 nanoparticles, following the equation:

$$U^{4+} + 4OH^- \leftrightarrow UO_2\downarrow + 2H_2O \quad (11)$$

2.5 Fluidized bed syntheses

Thermal denitration in a fluidized bed is another way to indirectly obtain UO_2 micro (and nano) particles. It involves spraying a concentrated solution of UNH on a bed of UO_3 at moderated temperatures (240–450°C) and fluidizing it with air or steam. The UO_3 produced nucleates on the existing UO_3 particles of the bed, enlarging their volume, or forming new particles. Afterward, thermal treatments can be used to convert the UO_3 to U_3O_8 and UO_2. This method uses less chemicals than the precipitation type of syntheses and allows the recuperation of the solvents but reported grain sizes are in the 100–500 μm [30], i.e., three orders of magnitude larger than the nanoparticles. However, it should be noted here that the equipment reviewed in most of the publications regarding fluidized beds are tuned to fabricate nuclear fuels. Under certain conditions of bed lengths, temperature, and solution feed speed, the authors reported the formation of "a very fine powder, not well suited to the subsequent powder handling" that is elutriated in the process [31]. This means that those grains smaller than some microns were separated by their different density, grain size and morphology in the vapor/gas stream, losing all information about the possible existence of nanoparticles. Thus, it is possible that nanoparticles would be obtained in fluidized bed denitration by tuning appropriate operative characteristic.

3. Radiation-assisted syntheses

This type of UO_2 nanoparticles syntheses focus on the reduction of U^{6+} to U^{4+} by some kind of radiation. The process is induced exposing an aqueous solution of U^{6+} and an organic precursor to beta particles or photons including gamma and X-rays.

When particles are used, they are typically 4.5–7 MeV electrons from particle accelerators [32–35]. For example, Roth et al. used a pulsed beam with a frequency of 12.5 Hz and 4 μs pulse duration with an average dose rate of 24 Gy/s. To get a dose of 15 kGy, 625 s of effective irradiation must be accumulated. Conversion efficiency of U^{6+} to U^{4+} was 95% after 15 kGy delivered dose. However, Pavelková et al. used doses up to 100 kGy of 4.5 MeV electrons. In the first case, the authors obtained a narrow size distribution of 22–35 nm nanoparticles and a BET surface area of 60–70 m^2/g [35]. In the second case, heat treatments were necessary to obtain well-developed nanocrystals with linear crystallite size 13–27 nm and specific surface area 10–46 m^2/g [32].

On the other hand, gamma-ray photons consist mainly in those from ^{60}Co radiation sources (two emissions of 1.17 and 1.33 MeV). Dose rates in the order of 198 Gy/h are delivered, and after 70 h of irradiation, 65% of conversion efficiency was obtained in the work of Roth et al. [35]. Nenoff et al. used also a 198 Gy/h setup, but irradiation times from 7 to 10 days. In these conditions, nanoparticles readily form in the solution [36]. After 7 days of irradiation time, Roth et al. obtained nanoparticles of around 80 nm [35], and Nenoff et al. found in fresh prepared solution 6 nm particles, while aging resulted in their agglomeration. In that work, the crystal phase was studied from the TEM diffraction pattern resulting in alpha (α)-U or orthorhombic U metal phase (space group Cmcm). These particles converted naturally to the fcc UO_2 crystal phase, when allowed to rest in air by some days [36].

Moreover, X-rays can be used also to generate nanoparticle precursors in the bulk of an uranyl nitrate solution, which after a thermal treatment below 600°C, transform to UO_2 nanoparticles. X-rays from medium pressure 140 W mercury lamps have been used for this purpose [32, 37, 38]. In medium-pressure mercury-vapor lamps, the lines from 200 to 600 nm are present, namely 253.7, 365.4, 404.7, 435.8, 546.1, and 578.2 nm. However in this case, the 253.7 nm line is the one of interest. Illumination times between 60 and 180 minutes were used to obtain the nanoparticle precursors. After that, a heat treatment under Ar:H_2 atmosphere at 550°C was done in order to form the UO_2 nanoparticles. A yield of 70% was obtained with this method. Nanoparticles obtained were monocrystals of 14.9 nm as determined by XRD spectra in accordance with TEM images and presented a specific surface area of 10.4 m^2/g.

3.1 Precursor formation: U^{6+} to U^{4+} reduction and polymerization

Regardless of the type of radiation used, U^{6+} is in the form of an uranyl nitrate $UO_2(NO_3)_2$ acidic aqueous solution and the uranyl concentrations used range from 10 to 50 mM. The chemical reduction reactions, which were presented by Rath et al., are described next [34]. The irradiation with electrons or photons produces water photolysis:

$$H_2O \xrightarrow{e^-, hv} H\bullet, OH\bullet, OH_2\bullet, H_2O_2, H_3O^+ \tag{12}$$

$$e^-_{aq} + H \rightarrow H\bullet \tag{13}$$

where H• is a reducing agent and OH• is an oxidizing radical. By adding an organic species, which commonly is a secondary alcohol, these species are scavenged, and a strongly reducing agent is produced (reactions Eqs. (14) and (15)). Propan-2-ol, for example, reacts with both H• and OH• forming a strongly reducing 1-hydroxy 2-propyl radical $H_3C{-}C\bullet OH\ CH_3$:

$$H_3C - CHOH - CH_3 + H\bullet \rightarrow H_3C - C\bullet OH - CH_3 + H_2 \tag{14}$$

$$H_3C - CHOH - CH_3 + OH\bullet \rightarrow H_3C - C\bullet OH\ CH3 + H_2O \tag{15}$$

Thus, in that milieu, the following reducing and polymerization reactions are possible:

$$UO_2^{2+} + e^-_{aq} \rightarrow UO_2^{1+} \tag{16}$$

$$UO_2^{2+} + H_3C - C\bullet OH - CH_3 \rightarrow UO_2^{+} + H_3C - CHOH - CH_3 + H^+ \tag{17}$$

$$UO_2^{1+} + H_3C - C\bullet OH - CH_3 \rightarrow UO_2 + H_3C - CO - CH_3 + H^+ \tag{18}$$

$$UO_2^{+} + nUO_2 \rightarrow (UO_2)\ \text{nanoparticle} \tag{19}$$

The following reaction (Eq. (20)) is also possible, which retires hydrated electrons from the solution, though ${NO_3}^{-2}$ anions finally convert to NO_3^-:

$$NO_3^- + e^-_{aq} \rightarrow NO_3^{-2} \tag{20}$$

During irradiation, there is ${UO_2}^{2+}$ consumption to form the ${UO_2}^{\circ}$ nanoparticles. The UV-visible absorption spectra of uranyl nitrate exhibit maxima at 427, 477, and 495 nm; the maxima gradually disappear during irradiation, due to the precipitation of the precursor. This is a usual way to follow the nanoparticle precursor formation kinetics during irradiation.

According to Rath et al., an induction time of 135 min after irradiation was necessary for the nanoparticles to form in the presence of 1% volume fraction of propan-2-ol and 50 kGy of absorbed dose [34]. The same work shows that this value depends on the scavenger concentration and the viscosity of the uranyl solution. Induction time also increased with the ethylene glycol concentration, which was used to obtain higher viscosity values.

4. Biologically assisted synthesis

Nanoparticles UO_2 can be obtained also by mediation of living organisms. *Shewanella* genus, for example, belongs to a well-known group of U^{6+} reducing bacteria. Within this group, anaerobic *Shewanella oneidensis MR-1* and *Shewanella putrefaciens CN32* species have been widely used by several authors to produce biogenic uraninite nanoparticles [39–42]. Other species studied were: *Desulfovibrio vulgaris* [43, 44], *Geobacter sulfurreducens* [45, 46], and *Anaeromyxobacter dehalogenans 2CP-C* [47].

One of the main interests related to this topic is the possible use of bacteria in reducing the environmental mobility of the U^{6+} ions by transforming them into U^{4+} species. Thus, there is a big effort in determining the factors that affect the physiological state of the microorganisms, which mediate the U^{6+} reduction, as well as in determining which geochemical and environmental conditions modify the nanobiogenic UO_2 surface reactivity [40] and redox potentials [42].

These experiments are conducted in a series of stages: the preparation of a background electrolyte, where the bacteria is allowed to live and growth, the cell cultivation, the U^{6+} bioreduction experiments and, finally, the determination of U^{4+} re-oxidation rates under different conditions. In natural environments, uranium might be present in different sites due to the geology of the area but also as a contaminant in soils, sediments, and groundwater [40]. So, on the one hand, the background electrolyte implies the preparation of buffered (6.8–8 pH range) artificial groundwater made of uranyl acetate in 1.2–4 mM concentrations and some organic additives as lactic acid and macronutrients for bacterial growth. On the other hand, cell suspensions are cultured aerobically at 30°C for 24 h; centrifuged, washed with an anaerobic buffer, and resuspended in an anaerobic solution. From this suspension, a portion is inoculated into the buffered, anaerobic uranyl-bearing solution to initiate uranium reduction. After a bioreduction essays, cell-uranium precipitates are pasteurized at 70°C to deactivate biological activity [40–42]. Burgos et al., for example, reported that it was challenging to determine what constitutes a single discrete particle in samples with thick uraninite coatings or large extracellular deposits but regardless of the bioreduction rate or the electrolyte used, identified a mean particle size structure of around 3 nm in TEM micrographs as well as with X-ray absorption fine structure spectroscopy (EXAFS) [39], while Singer et al. found stoichiometric uraninite with particle diameters of 5–10 nm by DRX [40].

Probably one of the most interesting results obtained by these authors was that the bioreduction rate is not the unique factor which controls the particle size of biogenic uraninite. Within the parameters that influence the obtaining of certain particle size, it can be include cell cultivation methods, metabolic state of cells, molecular-scale mechanisms of U^{6+} reduction, U^{4+} nucleation site, and cellular location of uraninite precipitates [39].

5. Conclusions

With the recent knowledge gained on nanoscience and nanomaterials, and the complex interaction that nanoparticles have in the environment, there is a new insight toward nanoparticles generated from the nuclear technology. It is a fact that long-term nuclear waste disposals and nuclear reactors are sources of UO_2 and actinide- and lanthanide-doped UO_2 nanoparticles. Therefore there is an effort to produce nanoparticles of these compositions to study not only their behavior in special physicochemical conditions but also their advantageous properties in the design of new fuel elements and processes.

There exist many ways to obtain nanoparticles of UO_2, but until now all of them start from a solution of U^{6+} and reduces it to U^{4+}. The way in which the nanoparticle is formed or the reduction is done differentiates one of the other processes. In the precipitation routes, the pH generates nanoparticles of U^{6+} salts that after intermediate- to high-temperature thermal treatments in reducing conditions convert to micrometer agglomerates of UO_2 nanoparticles of 80–120 nm and fcc crystal phase. Other chemical routes use U^{6+} organic salts in an organic solvent as dibenzyl ether, with amines and organic acids as stabilization agents, to induce the precipitation of non-agglomerated and highly crystalline UO_2 nanoparticles of less than 5 nm during a low-temperature thermal treatment. In the sol-gel type syntheses, nanoparticles with U^{6+} are generated in the continuous solid phase, sometimes mediated by the addition of organic molecules. The gel is dried after and reduced to obtain micrometer-sized agglomerates of UO_2 nanoparticles of around 90 nm crystallite size. In the electrochemical-assisted syntheses, electrons are directly supplied at the cathode to the uranyl solution to reduce the uranium ions to U^{4+}, which precipitates as moderately agglomerated powders of 53 nm formed by 4–14 nm crystal size UO_2 nanoparticles. The processes assisted by radiation consist in generating strongly reducing organic agents by irradiating a secondary alcohol with electrons or photons. These species reduce the U^{6+} to U^{4+} in the solution forming UO_2, which aggregates in crystalline nanoparticles. In case of electron irradiation, small particles with a narrow size distribution (22–35 nm) were obtained, while for gamma irradiation 3.5–5 nm particles were formed. In case of X-rays photons, the product obtained are precursors of nanoparticles and need a subsequent intermediate-temperature thermal treatment to definitely form the UO_2 nanoparticles with 3–15 nm and fcc crystal phase.

Acknowledgements

The authors thank the INN for the financial support to publish this chapter.

Author details

Analía Leticia Soldati[1,2]*, Diana Carolina Lago[1] and Miguel Oscar Prado[1,2]

1 Nuclear Materials Department, Bariloche Atomic Center, National Atomic Energy Commission (CNEA), San Carlos de Bariloche, Argentina

2 Institute of Nanoscience and Nanotechnology (INN), National Commission of Science and Technology Research (CONICET), San Carlos de Bariloche, Argentina

*Address all correspondence to: asoldati@cab.cnea.gov.ar

References

[1] Hochella MF et al. Natural, incidental, and engineered nanomaterials and their impacts on the Earth system. Science. 2019;**363**(6434): eaau8299

[2] Spino J, Vennix K, Coquerelle M. Detailed characterisation of the rim microstructure in PWR fuels in the burn-up range 40-67 GWd/tM. Journal of Nuclear Materials. 1996;**231**(3): 179-190

[3] Spino J et al. Bulk-nanocrystalline oxide nuclear fuels—An innovative material option for increasing fission gas retention, plasticity and radiation-tolerance. Journal of Nuclear Materials. 2012;**422**(1–3):27-44

[4] Soldati AL et al. Synthesis and characterization of Gd_2O_3 doped UO_2 nanoparticles. Journal of Nuclear Materials. 2016;**479**:436-446

[5] Cappia F et al. Laser melting of nano-crystalline uranium dioxide. Progress in Nuclear Energy. 2014;**72**(0):11-16

[6] Jovani-Abril R et al. Synthesis of nc-UO_2 by controlled precipitation in aqueous phase. Journal of Nuclear Materials. 2016;**477**:298-304

[7] Nenoff TM et al. Synthesis and low temperature in situ sintering of uranium oxide nanoparticles. Chemistry of Materials. 2011;**23**(23):5185-5190

[8] Kalmykov SN, Denecke M. Actinide Nanoparticle Research. Berlin Heidelberg: Springer-Verlag; 2011

[9] Neill TS et al. Silicate stabilisation of colloidal UO_2 produced by uranium metal corrosion. Journal of Nuclear Materials. 2019;**526**:151751

[10] Fuller EL et al. Uranium oxidation: Characterization of oxides formed by reaction with water by infrared and sorption analyses. Journal of Nuclear Materials. 1984;**120**(2):174-194

[11] Kaminski MD et al. Colloids from the aqueous corrosion of uranium nuclear fuel. Journal of Nuclear Materials. 2005;**347**(1):77-87

[12] Weber WG et al. Chapter 2: Preparation of uranium dioxide. In: Belle J, editor. Uranium Dioxide: Properties and Nuclear Applications. Washington: Naval Reactors, Division of Reactor Development USA Energy Commission; 1921. pp. 35-64

[13] Menghini J et al. Mixed oxide pellets obtention by the "Reverse Strike" co-precipitation method. Session 1: Optimization of fuel fabrication technology practices and modelling. In: Bibilashvili Y, Dörr W, editors. IAEA Technical Meeting. Brussel, Belgien: IAEA; 2004. pp. 31-44

[14] Riella HG et al. UO_2-Gd_2O_3 solid solution formation from wet and dry processes. Journal of Nuclear Materials. 1991;**178**(2–3):204-211

[15] Fukushima S et al. The effect of gadolinium content on the thermal conductivity of near-stoichiometric (U, Gd)O_2 solid solutions. Journal of Nuclear Materials. 1982;**105**(2):201-210

[16] Miyake C, Kanamaru M, Imoto S. Microcharacterization of gadolinium in U1—xGdxO_2 by means of electron spin resonance. Journal of Nuclear Materials. 1986;**137**(3):256-260

[17] Durazzo M et al. Phase studies in the UO_2-Gd_2O_3 system. Journal of Nuclear Materials. 2010;**400**(3):183-188

[18] Wu H, Yang Y, Cao YC. Synthesis of colloidal uranium-dioxide nanocrystals. Journal of the American Chemical Society. 2006;**128**(51):16522-16523

[19] Hudry D et al. Non-aqueous synthesis of isotropic and anisotropic actinide oxide nanocrystals. Chemistry—A European Journal. 2012; **18**:8283-8287

[20] Hu S et al. Nanocrystals of uranium oxide: Controlled synthesis and enhanced electrochemical performance of hydrogen evolution by Ce doping. Small. 2015;**11**(22):2624-2630

[21] Wang Q et al. Synthesis of uranium oxide nanoparticles and their catalytic performance for benyl alcohol conversion to benzaldehyde. Journal of Materials Chemistry. 2008;**18**:1146-1152

[22] Tyrpekl V et al. Low temperature decomposition of U(IV) and Th(IV) oxalates to nanograined oxide powders. Journal of Nuclear Materials. 2015;**460**: 200-208

[23] Brinker CJ, Scherer GW. Chapter 1—Introduction. In: Brinker CJ, Scherer GW, editors. Sol-Gel Science. San Diego: Academic Press; 1990. pp. xvi-18

[24] Ziouane Y et al. Effect of the microstructural morphology on UO_2 powders. Procedia Chemistry. 2016;**21**: 319-325

[25] Daniels H et al. Fabrication of oxidic uranium-neodymium microspheres by internal gelation. Progress in Nuclear Energy. 2012;**57**:106-110

[26] Schreinemachers C et al. Characterization of uranium neodymium oxide microspheres synthesized by internal gelation. Progress in Nuclear Energy. 2014;**72**: 17-21

[27] Gündüz G, Önal I, Durmazuçar HH. Pore size distributions in uranium dioxide and uranium dioxide-gadolinium oxide fuel kernel produced by sol-gel technique. Journal of Nuclear Materials. 1991;**178**(2–3):212-216

[28] Leblanc M et al. Actinide mixed oxide conversion by advanced thermal denitration route. Journal of Nuclear Materials. 2019;**519**:157-165

[29] Rousseau G et al. Synthesis and characterization of nanometric powders of UO_{2+x}, $(Th,U)O_{2+x}$ and $(La,U)O_{2+x}$. Journal of Solid State Chemistry. 2009; **182**(10):2591-2597

[30] Fane A, Charlton B, Alfredson P. The Thermal Denitration of Uranyl Nitrate in a Fluidised Bed Reactor. Sidney: ANSTO; 1974

[31] Harrington CD, Ruehle AE. Uranium Production Technology. Princeton, New Jersey: D. Van Nostrand Company Inc.; 1959

[32] Pavelková T et al. E-beam and UV induced fabrication of CeO_2, Eu_2O_3 and their mixed oxides with UO_2. Radiation Physics and Chemistry. 2016;**124**: 252-257

[33] Rath MC, Naik DB. Post-irradiation induction time in the radiolytic synthesis of UO_2 nanoparticles in aqueous solutions. Journal of Nuclear Materials. 2014;**454**(1):54-59

[34] Rath MC, Naik DB, Sarkar SK. Reversible growth of UO_2 nanoparticles in aqueous solutions through 7.0 MeV electron beam irradiation. Journal of Nuclear Materials. 2013;**438**(1–3):26-31

[35] Roth O, Hasselberg H, Jonsson M. Radiation chemical synthesis and characterization of UO_2 nanoparticles. Journal of Nuclear Materials. 2009; **383**(3):231-236

[36] Nenoff TM et al. Formation of uranium based nanoparticles via gamma-irradiation. Journal of Nuclear Materials. 2013;**442**(1–3):162-167

[37] Pavelková T et al. Preparation of UO_2, ThO_2 and $(Th,U)O_2$ pellets

from photochemically-prepared nano-powders. Journal of Nuclear Materials. 2016;**469**:57-61

[38] Pavelková T, Čuba V, Šebesta F. Photo-induced low temperature synthesis of nanocrystalline UO_2, ThO_2 and mixed UO_2-ThO_2 oxides. Journal of Nuclear Materials. 2013;**442**(1):29-32

[39] Burgos WD et al. Characterization of uraninite nanoparticles produced by *Shewanella oneidensis* MR-1. Geochimica et Cosmochimica Acta. 2008;**72**(20): 4901-4915

[40] Singer DM, Farges F, Brown GE Jr. Biogenic nanoparticulate UO_2: Synthesis, characterization, and factors affecting surface reactivity. Geochimica et Cosmochimica Acta. 2009;**73**(12): 3593-3611

[41] Schofield EJ et al. Structure of biogenic uraninite produced by *Shewanella oneidensis* strain MR-1. Environmental Science & Technology. 2008;**42**(21):7898-7904

[42] Ulrich K-U et al. Comparative dissolution kinetics of biogenic and chemogenic uraninite under oxidizing conditions in the presence of carbonate. Geochimica et Cosmochimica Acta. 2009;**73**(20):6065-6083

[43] Zhou C et al. Growth of *Desulfovibrio vulgaris* when respiring U(VI) and characterization of biogenic uraninite. Environmental Science & Technology. 2014;**48**(12):6928-6937

[44] Lovley DR et al. Reduction of uranium by cytochrome c3 of *Desulfovibrio vulgaris*. Applied and Environmental Microbiology. 1993; **59**(11):3572-3576

[45] Sharp JO et al. Uranyl reduction by *Geobacter sulfurreducens* in the presence or absence of iron. In: Merkel BJ, Hasche-Berger A, editors. Uranium, Mining and Hydrogeology. Berlin, Heidelberg: Springer; 2008

[46] Bargar JR, et al. Coupled biogeochemical processes governing the stability of bacteriogenic UO_{2+x}. In: 3rd Annual DOE-ERSP PI Meeting. Lansdowne, Virginia; 2008

[47] Wu Q, Sanford RA, Löffler FE. Uranium(VI) reduction by *Anaeromyxobacter dehalogenans* strain 2CP-C. Applied and Environmental Microbiology. 2006;**72**(5):3608-3614

Section 3

Relevant Measurement Methods

Chapter 3

Exergy: Mechanical Nuclear Physics Measures Pressure, Viscosity and X-Ray Resonance in K-Shell in a Classical Way

Edward Henry Jimenez

Abstract

First, the liquid drop model assumes *a priori*; to the atomic nucleus composed of protons and neutrons, as an incompressible nuclear fluid that should comply with the Navier–Stokes 3D equations (N-S3D). Conjecture, not yet proven, however, this model has successfully predicted the binding energy of the nuclei. Second, the calculation of nuclear pressure $(p_0 \in 1.42, 1.94]10^{32} Pa)$ and average viscosity $(\eta = 1.71 \times 10^{24} [fm^2/s])$, as a function of the nuclear decay constant $k = \frac{p_0}{2\eta} = \frac{1}{T_{1/2}}$, not only complements the information from the National Nuclear Data Center, but also presents an analytical solution of (N- S3D). Third, the solution of (N-S3D) is a Fermi Dirac generalized probability function $P(x, y, z, t) = \frac{1}{1+e^{\frac{p_0}{2\eta}t-\mu\left(x^2+y^2+z^2\right)^{1/2}}}$, Fourth, the parameter μ has a correspondence with the Yukawa potential coefficient $\mu = \alpha m = 1/r$, Fifth, using low energy X-rays we visualize and measure parameters of the nuclear surface (proton radio) giving rise to the femtoscope. Finally, we obtain that the pressure of the proton is 8.14 times greater than the pressure of the neutron, and 1000 times greater than the pressure of the atomic nucleus. Analyzed data were isotopes: $9 \leq Z \leq 92$ and $9 \leq N \leq 200$.

Keywords: femtoscope, Navier Stokes 3D, nuclear viscosity, minimum entropy

1. Introduction

Neutron stars are among the densest known objects in the universe, withstanding pressures of the order of $10^{34} Pa$. However, it turns out that protons [1], the fundamental particles that make up most of the visible matter in the universe, contain pressures 10 times greater, [2, 3] $10^{35} Pa$. This has been verified from two perspectives at the Jefferson Laboratory, MIT [1–5]. High-energy physics continue to guide the study of the mechanical properties of the subatomic world.

Viscosity is a characteristic physical property of all fluids, which emerges from collisions between fluid particles moving at different speeds, causing resistance to their movement (**Figure 1**). When a fluid is forced to move by a closed surface, similar to the atomic nucleus, the particles that make up the fluid move slower in the center and faster on the walls of the sphere. Therefore, a shear stress (such as a pressure difference) is necessary to overcome the friction resistance between the

IntechOpen

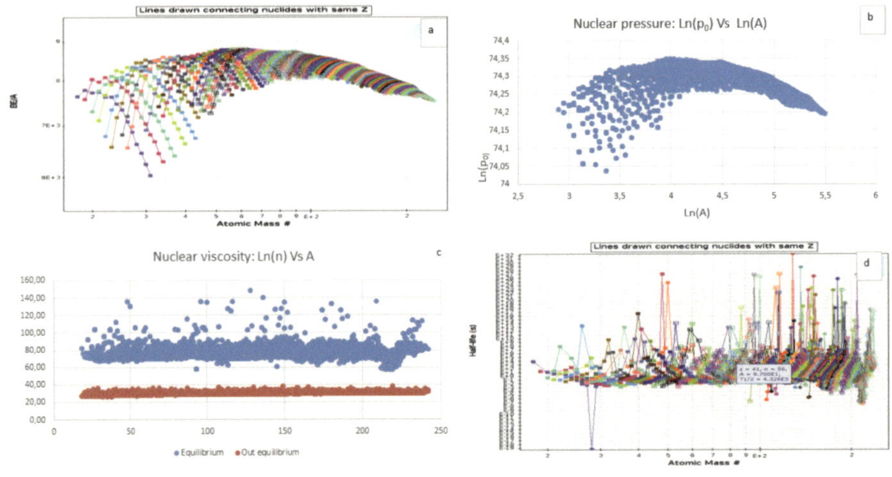

Figure 1.
Obtaining nuclear viscosity and nuclear pressure from the speed of the neutron particles in the disintegration of chemical element. ***Figure 1a.*** *indicates that the BE/A ratio is proportional to the nuclear pressure,* p_o *represented in* ***Figure 1b.*** ***Figure 1c*** *is the graph of the viscosity in equilibrium and out of equilibrium, the viscosity in equilibrium is greater than the viscosity at the moment of nuclear decay.* ***Figure 1d*** *is the average half-time of each isotope,* $9 \leq Z \leq 92$ *and* $9 \leq N \leq 200$.

layers of the nuclear fluid. For the same radial velocity profile, the required tension is proportional to the viscosity of the nuclear fluid or its composition given by (Z, N) [6–8].

Radioactive decay is a stochastic process, at the level of individual atoms. According to quantum theory, it is impossible to predict when a particular atom will decay, regardless of how long the atom has existed. However, for a collection of atoms of the same type [8], the expected decay rate is characterized in terms of its decay constant $k = \frac{1}{T_{1/2}}$. The half-lives of radioactive atoms do not have a known upper limit, since it covers a time range of more than 55 orders of magnitude, from almost instantaneous to much longer than the age of the universe [8, 9].

Other characteristics of the proton such as its size have been studied in many institutes such as Max Plank, where it has been measured with high precision ranges $\left(r_p = 0.84184(67)fm\right)$, providing new research methods [10, 11].

The atomic nucleus is an incompressible fluid, justified by the formula of the nuclear radius, $R = 1.2A^{\frac{1}{3}}$, where it is evident that the volume of the atomic nucleus changes linearly with $A = Z + N$, giving a density constant [11]. All incompressible fluid and especially the atomic nucleus comply with the Navier Stokes equations. We present a rigorous demonstration on the incomprehensibility of the atomic nucleus, which allows to write explicitly the form of the nuclear force $\mathbf{F}_N = -\frac{g\mu^2}{8\pi}(A-1)P(1-P)\nabla\mathbf{r}$, which facilitates the understanding of nuclear decay.

The Navier Stokes equations are a problem of the millennium [12, 13], that has not been resolved yet in a generalized manner. We present a solution that logically meets all the requirements established by the Clay Foundation [14, 15]. This solution coherently explains the incompressible nuclear fluid and allows calculations of the nuclear viscosity and nuclear pressure [1, 2].

The alpha particle is one of the most stable. Therefore it is believed that it can exist as such in the heavy core structure. The kinetic energy typical of the alpha particles resulting from the decay is in the order of 5 MeV.

For our demonstrations, we will use strictly the scheme presented by Fefferman in *http://www.claymath.org/millennium-problems* [13, 14], where six demonstrations are required to accept as valid a solution to the Navier–Stokes 3D Equation [16–18]. An understanding of the mechanics of the atomic nucleus cannot do without fluid equations.

2. Model

The velocity defined as $\mathbf{u} = -2\nu \frac{\nabla P}{P}$, with a radius noted as $r = (x^2 + y^2 + z^2)^{1/2}$ where $P(x,y,z,t)$ is the logistic probability function $P(x,y,z,t) = \frac{1}{1+e^{kt-\mu r}}$, and the expected value $E(r|r \geq 0) < C$ exist. The term P is defined in $((x,y,z) \in \mathbb{R}^3, t \geq 0)$, where constants $k > 0, \mu > 0$ and $P(x,y,x,t)$ is the general solution of the Navier–Stokes 3D equation, which has to satisfy the conditions (1) and (2), allowing us to analyze the dynamics of an incompressible fluid [12–14].

$$\frac{\partial \mathbf{u}}{\partial t} + (\mathbf{u}.\nabla)\mathbf{u} = \nu \nabla^2 \mathbf{u} - \frac{\nabla p}{\rho_0} \qquad ((x,y,z) \in \mathbb{R}^3, t \geq 0) \tag{1}$$

With, $\mathbf{u} \in \mathbb{R}^3$ an known velocity vector, ρ_0 constant density of fluid, η dynamic viscosity, ν cinematic viscosity, and pressure $p = p_0 P$ in $((x,y,z) \in \mathbb{R}^3, t \geq 0)$.

Where velocity and pressure are depending of r and t. We will write the condition of incompressibility.

$$\nabla.\mathbf{u} = 0 \qquad ((x,y,z) \in \mathbb{R}^3, t \geq 0) \tag{2}$$

The initial conditions of fluid movement $\mathbf{u}^0(x,y,z)$, are determined for $t = 0$. Where speed $\mathbf{u}^0$ must be C^∞ divergence-free vector.

$$\mathbf{u}(x,y,z,0) = \mathbf{u}^0(x,y,z) \qquad ((x,y,z) \in \mathbb{R}^3) \tag{3}$$

For physically reasonable solutions, we make sure $\mathbf{u}(x,y,z,t)$ does not grow large as $r \to \infty$. We will restrict attention to initial conditions $\mathbf{u}^0$ that satisfy.

$$\left|\partial_x^\alpha \mathbf{u}^0\right| \leq C_{\alpha K}(1+r)^{-K} \qquad \text{on } \mathbb{R}^3 \text{ for any } \alpha \text{ and } K \tag{4}$$

The Clay Institute accepts a physically reasonable solution of (1), (2) and (3), only if it satisfies:

$$p, \mathbf{u} \in C^\infty\left(\mathbb{R}^3 \times 0, \infty)\right) \tag{5}$$

and the finite energy condition [14–16].

$$\int_{\mathbb{R}^3} |\mathbf{u}(x,y,z,t)|^2 dxdydz \leq C \qquad \text{for all } t \geq 0 \qquad (\text{bounded energy}). \tag{6}$$

The problems of Mathematical Physics are solved by the Nature, guiding the understanding, the scope, the limitations and the complementary theories. These guidelines of this research were: the probabilistic elements of Quantum Mechanics, the De Broglie equation and the Heisenberg Uncertainty principle.

2.1 Definitions

Nuclear reaction velocity coefficient.

We will use an equation analogous to concentration equation of Physical Chemistry $C = C_0 e^{kt}$, where $k = \frac{p_0}{2\eta}$, is velocity coefficient, p_0 is the initial pressure of our fluid, η the dynamic viscosity and C_0 the initial concentration of energetic fluid molecules.

It is evident that, in equilibrium state we can write $\mu r = kt$, however, the Navier–Stokes equation precisely measures the behavior of the fluids out of equilibrium, so that: $\mu r \neq kt$.

Fortunately, there is a single solution for out-of-equilibrium fluids, using the fixed-point theorem for implicit functions, $\frac{1}{1+e^{kt-\mu r}} = \frac{2}{\mu r}$, the proof is proved in *Theorem* 1.

Attenuation coefficient.

We will use the known attenuation formula of an incident flux I_0, for which $I = I_0 e^{-\mu r}$. Where, I_0 initial flux and μ attenuation coefficient of energetic molecules that enter into interaction and/or resonance with the target molecules, transmitting or capturing the maximum amount of energy [5].

Dimensional analysis and fluid elements.

We will define the respective dimensional units of each one of variables and physical constants that appear in the solution of the Navier–Stokes 3D equation [12–14, 16].

Nuclear decay $N(t) = N_0 e^{-kt}$. Where $k = \frac{1}{T_{1/2}}$ [1/s], the velocity coefficient, and $T_{1/2}$ ground state half-life.

Kinematic viscosity $\nu = \frac{\eta}{\rho_0}$, $\left[\frac{m^2}{s}\right]$.

Dynamic viscosity η, [pa.s], where pa represents pascal pressure unit.

Initial Pressure of out of equilibrium. p_0, [pa].

Fluid density ρ_0, $\left[\frac{kg}{m^3}\right]$, where *kg* is kilogram and m^3 cubic meters.

Logistic probability function, $P(x,y,z,t) = \frac{1}{1+e^{kt-\mu r}}$, it is a real number $0 \leq P \leq 1$.

Equilibrium condition, $r = \frac{k}{\mu} t = \frac{p_0}{2\rho_0 \nu \mu} t = |\mathbf{u}_e| t$, [m].

Fluid velocity in equilibrium, $|\mathbf{u}_e|$, $[m/s]$. Protons, neutrons and alpha particles are the elements of the fluid.

Fluid field velocity out of equilibrium, $\mathbf{u} = -2\nu\mu(1-P)\nabla r$. [m/s]. All nuclear decay is a process out of nuclear equilibrium.

Position, $r = (x^2 + y^2 + z^2)^{1/2}$, [m].

Attenuation coefficient, μ, [1/m].

Growth coefficient, $k = \frac{p_0}{2\rho_0 \nu} = \frac{p_0}{2\eta}$, [1/s].

Concentration $C = C_0 \frac{1-P}{P}$.

Theorem 1 *The velocity of the fluid is given by* $\mathbf{u} = -2\nu \frac{\nabla P}{P}$, *where* $P(x,y,z,t)$ *is the logistic probability function* $P(x,y,x,t) = \frac{1}{1+e^{kt-\mu r}}$, *and p pressure such that* $p = p_0 P$, *both defined on* $((x,y,z) \in \mathbb{R}^3, t \geq 0)$. *The function P is the general solution of the Navier Stokes equations, which satisfies conditions (1) and (2).*

Proof. To verify condition (2), $\nabla.\mathbf{u} = 0$, we must calculate the gradients and laplacians of the radius. $\nabla r = \left(\frac{x}{r}, \frac{y}{r}, \frac{z}{r}\right)$, and $\nabla^2 r = \nabla.\nabla r = \frac{(y^2+z^2)+(x^2+z^2)+(x^2+y^2)}{(x^2+y^2+z^2)^{3/2}} = \frac{2}{r}$.

$$\nabla.u = -2\nu\nabla.\frac{\nabla P}{P} = -2\nu\mu\nabla((1-P)\nabla r) \tag{7}$$

Replacing the respective values for the terms: $\nabla^2 r$ and $|\nabla r|^2$ in the Eq. (7).

$$\begin{aligned}\nabla.u &= -2\nu\mu\nabla((1-P)\nabla r)\\ &= -2\nu\mu\nabla((1-P)\nabla r)\\ &= -2\nu\mu\left[-\mu\left(P-P^2\right)|\nabla r|^2+(1-P)\nabla^2 r\right]\end{aligned} \tag{8}$$

Where the gradient modulus of $\nabla P = \mu\left(P-P^2\right)\nabla r$, has the form $|\nabla P|^2 = \mu^2\left(P-P^2\right)^2|\nabla r|^2 = \mu^2\left(P-P^2\right)^2$.

$$\nabla.u = -2\nu\mu(1-P)\left[-\mu P+\frac{2}{r}\right]=0 \tag{9}$$

Simplifying for $(1-P)\neq 0$, we obtain the main result of this paper, which represents a fixed point of an implicit function $f(t,r)$ where $f(t,r) = P-\frac{2}{\mu r}=0$. In Nuclear Physics, $r_0 < r < 1.2A^{1/3}$.

$$P=\frac{1}{1+e^{kt-\mu\left(x^2+y^2+z^2\right)^{1/2}}}=\frac{2}{\mu\left(x^2+y^2+z^2\right)^{1/2}} \qquad \left((x,y,z)\in\mathbb{R}^3, t\geq 0\right) \tag{10}$$

Eq. (10) has a solution according to the fixed-point theorem of an implicit function, and it is a solution to the Navier Stokes stationary equations, which are summarized in: $\nabla^2 P = \frac{2}{\mu}\nabla^2\left(\frac{1}{r}\right)=0$. Furthermore, it is the typical solution of the Laplace equation for the pressure of the fluid $\nabla^2 p = p_0\nabla^2 P = 0$. Kerson Huang (1987).

To this point, we need to verify that Eq. (10) is also a solution of requirement (1), $\frac{\partial \mathbf{u}}{\partial t}+(\mathbf{u}.\nabla)\mathbf{u}=\nu\nabla^2\mathbf{u}-\frac{\nabla p}{\rho_0}$. We will do the equivalence $\mathbf{u}=\nabla\theta$ after we replace in Eq. (1). Taking into account that $\theta=-2\nu\ln(P)$, and that $\nabla\theta$ is irrotational, $\nabla\times\nabla\theta=0$, we have: $(\mathbf{u}.\nabla)\mathbf{u}=(\nabla\theta.\nabla)\nabla\theta=\frac{1}{2}\nabla(\nabla\theta.\nabla\theta)-\nabla\theta\times(\nabla\times\nabla\theta)=\frac{1}{2}\nabla(\nabla\theta.\nabla\theta)$, and $\nabla^2\mathbf{u}=\nabla(\nabla.\mathbf{u})-\nabla\times(\nabla\times\mathbf{u})=\nabla(\nabla.\nabla\theta)-\nabla\times(\nabla\times\nabla\theta)=\nabla\left(\nabla^2\theta\right)$. Simplifying terms in order to replace these results in Eq. (1) we obtain

$$(\mathbf{u}.\nabla)\mathbf{u}=\frac{1}{2}\nabla(\nabla\theta.\nabla\theta)=2\nu^2\nabla\left(\frac{|\nabla P|^2}{P^2}\right)$$

$$\begin{aligned}\nabla^2\mathbf{u} &= \nabla(\nabla.\mathbf{u})=\nabla\left(\nabla^2\theta\right)=0\\ &= -2\nu\nabla\left(\frac{|\nabla P|^2}{P^2}-\frac{\nabla^2 P}{P}\right)=0\end{aligned}$$

The explicit form of velocity is $\mathbf{u}=-2\mu\nu(1-P)\nabla r$. Next, we need the partial derivative $\frac{\partial \mathbf{u}}{\partial t}$

$$\frac{\partial \mathbf{u}}{\partial t}=-2\mu\nu kP(1-P)\nabla r,$$

$$-\frac{\nabla p}{\rho_0}=-\frac{\mu p_0}{\rho_0}P(1-P)\nabla r.$$

After replacing the last four results $(\mathbf{u}.\nabla)\mathbf{u}$, $\nabla^2\mathbf{u}$, $\frac{\partial \mathbf{u}}{\partial t}$ and $-\frac{\nabla p}{\rho_0}$ in Eq. (1) we obtain (11).

$$-2\mu\nu kP(1-P)\nabla r = 2\nu^2\nabla\left(\frac{|\nabla P|^2}{P^2}\right) - \frac{\mu p_0}{\rho_0}P(1-P)\nabla r. \quad (11)$$

The Eq. (11) is equivalent to Eq. (1). After obtaining the term $\frac{|\nabla P|^2}{P^2}$ from the incompressibility equation $\nabla(\nabla^2\theta) = -2\nu\nabla\left(-\frac{|\nabla P|^2}{P^2} + \frac{\nabla^2 P}{P}\right) = 0$ and replacing in Eq. (11).

$$-2\mu\nu kP(1-P)\nabla r = 2\nu^2\nabla\left(\frac{\nabla^2 P}{P}\right) - \frac{\mu p_0}{\rho_0}P(1-P)\nabla r. \quad (12)$$

Eq. (10) simultaneously fulfills requirements (1) expressed by Eq. (12) and requirement (2) expressed by Eq. (7), for a constant $k = \frac{p_0}{2\rho_0\nu} = \frac{p_0}{2\eta}$. Moreover, according to Eq. (10), the probability $P = \frac{2}{\mu r}$ which allows the Laplace equation to be satisfied: $\nabla^2 P = \frac{2}{\mu}\nabla^2\left(\frac{1}{r}\right) = 0$. In other words, the Navier–Stokes 3D equation system is solved. ■

Implicit Function.

An implicit function defined as (10), $f(t,r) = \frac{1}{1+e^{kt-\mu r}} - \frac{2}{\mu r} = 0$ has a fixed point (t,r) of $R = \{(t,r)|0 < a \le t \le b, 0 < r < +\infty\}$, where m and M are constants, such as: $m \le M$. Knowing that the partial derivative exists: $\partial_r f(t,r) = \nu P(1-P) + \frac{2}{\mu r^2}$ we can assume that: $0 < m \le \partial_r f(t,r) \le M$. If, in addition, for each continuous function φ in $[a,b]$ the composite function $g(t) = f(t,\varphi(t))$ is continuous in $[a,b]$, then there is one and only one function: $r = \varphi(t)$ continuous in $[a,b]$, such that $f[t,\varphi(t)] = 0$ for all t in $[a,b]$.

Theorem 2 *An implicit function defined as (10)* $f(t,r) = \frac{1}{1+e^{kt-\mu r}} - \frac{2}{\mu r} = 0$ *has a fixed point* (t,r) *of* $R = \{(t,r)|0 < a \le t \le b, 0 < r < +\infty\}$. *In this way, the requirements (1) and (2) are fulfilled.*

Proof. Let C be the linear space of continuous functions in $[a,b]$, and define an operator $T : C \to C$ by the equation:

$$T\varphi(t) = \varphi(t) - \frac{1}{M}f[t,\varphi(t)].$$

Then we prove that T is a contraction operator, so it has a unique fixed point $r = \varphi(t)$ in C. Let us construct the following distance.

$$T\varphi(t) - T\psi(t) = \varphi(t) - \psi(t) - \frac{f[t,\varphi(t)] - f[t,\psi(t)]}{M}.$$

Using the mean value theorem for derivation, we have

$$f[t,\varphi(t)] - f[t,\psi(t)] = \partial_\phi f(t,z(t))[\varphi(t) - \psi(t)].$$

Where $\phi(t)$ is situated between $\varphi(t)$ and $\psi(t)$. Therefore, the distance equation can be written as:

$$T\varphi(t) - T\psi(t) = [\varphi(t) - \psi(t)]\left[1 - \frac{\partial_\phi f(t,z(t))}{M}\right]$$

Using the hypothesis $0 < m \le \partial_r f(t,r) \le M$ we arrive at the following result:

$$0 \le 1 - \frac{\partial_\phi f(t,\phi(t))}{M} \le 1 - \frac{m}{M},$$

with which we can write the following inequality:

$$|T\varphi(t) - T\psi(t)| = |\varphi(t) - \psi(t)|\left(1 - \frac{m}{M}\right) \leq \alpha\|\varphi - \psi\|. \tag{13}$$

Where $\alpha = \left(1 - \frac{m}{M}\right)$. Since $0 < m \leq M$, we have $0 \leq \alpha < 1$. The above inequality is valid for all t of $[a, b]$. Where T is a contraction operator and the proof is complete, since for every contraction operator $T : C \to C$ there exists one and only one continuous function φ in C, such that $T(\varphi) = \varphi$. Using Eq. (10), which represents the fundamental solution of the Navier–Stokes 3D equation, we verify Eq. (2), which represents the second of the six requirements of an acceptable solution. ■

Proposition 3 *Requirement (3). The initial velocity can be obtained from:* $\mathbf{u}(x, y, z, 0) = -2\nu\frac{\nabla P}{P}$, *where each of the components* u_x, u_y *and* u_z *are infinitely derivable.*

$$\mathbf{u}(x, y, z, 0) = \mathbf{u}^0(x, y, z) = -2\nu\mu(1 - P_0)\left(\frac{x}{r}, \frac{y}{r}, \frac{z}{r}\right) \qquad \left((x, y, z) \in \mathbb{R}^3\right)$$

$$P_0 = \frac{1}{1 + e^{-\mu r_0}} \tag{14}$$

Proof. Taking the partial derivatives of $\partial_x^n\left(\frac{x}{r}\right)$, $\partial_y^n\left(\frac{x}{r}\right)$ and $\partial_z^n\left(\frac{x}{r}\right)$.

$$\begin{aligned} \partial_x^n\left(\frac{x}{r}\right) &= n\partial_x^{n-1}\left(\frac{1}{r}\right) + x\partial_x^n\left(\frac{1}{r}\right) \\ \partial_y^n\left(\frac{x}{r}\right) &= n\partial_y^{n-1}\left(\frac{1}{r}\right) + y\partial_y^n\left(\frac{1}{r}\right) \\ \partial_z^n\left(\frac{x}{r}\right) &= n\partial_z^{n-1}\left(\frac{1}{r}\right) + z\partial_z^n\left(\frac{1}{r}\right) \end{aligned} \tag{15}$$

Recalling the derivatives of special functions (Legendre), it is verified that there exists the derivative C^∞.

$$\begin{aligned} \partial_x^n\left(\frac{1}{r}\right) &= (-1)^n n!\left(x^2 + y^2 + z^2\right)^{-\frac{(n+1)}{2}} P_n\left(\frac{x}{(x^2 + y^2 + z^2)^{1/2}}\right) \\ \partial_y^n\left(\frac{1}{r}\right) &= (-1)^n n!\left(x^2 + y^2 + z^2\right)^{-\frac{(n+1)}{2}} P_n\left(\frac{y}{(x^2 + y^2 + z^2)^{1/2}}\right) \\ \partial_z^n\left(\frac{1}{r}\right) &= (-1)^n n!\left(x^2 + y^2 + z^2\right)^{-\frac{(n+1)}{2}} P_n\left(\frac{z}{(x^2 + y^2 + z^2)^{1/2}}\right) \end{aligned} \tag{16}$$

■

Physically, this solution is valid for the initial velocity, indicated by Eq. (4), where the components of the initial velocity are infinitely differentiable, and make it possible to guarantee that the velocity of the fluid is zero when $r \to \infty$ [6–9].

Proposition 4 *Requirement (4). Using the initial velocity of a moving fluid given by* $\mathbf{u}(x, y, z, 0) = \mathbf{u}^0(x, y, z) = -2\nu\mu(1 - P_0)\left(\frac{x}{r}, \frac{y}{r}, \frac{z}{r}\right)$, *it is evident that*

$$\left|\partial_x^\alpha \mathbf{u}^0\right| \leq C_{\alpha K}(1 + r)^{-K} \qquad \textit{on } \mathbb{R}^3 \textit{ for any } \alpha \textit{ and } K$$

Proof. Using the initial velocity of a moving fluid given by $\mathbf{u}^0(x, y, z) = -2\nu\mu(1 - P_0)\left(\frac{x}{r}, \frac{y}{r}, \frac{z}{r}\right)$, we can find each of the components: $\partial_x^\alpha u_x^0$, $\partial_y^\alpha u_y^0$ and $\partial_z^\alpha u_z^0$.

$$\left(\partial_x^\alpha \frac{x}{r}\right)^2 = \left(\alpha\partial_x^{\alpha-1}\left(\frac{1}{r}\right) + x\partial_x^\alpha\left(\frac{1}{r}\right)\right)\left(\alpha\partial_x^{\alpha-1}\left(\frac{1}{r}\right) + x\partial_x^\alpha\left(\frac{1}{r}\right)\right)$$

For the three components x, y, z the results of the partial derivatives are as follows:

$$\begin{aligned}
\left(\partial_x^\alpha \frac{x}{r}\right)^2 &= \alpha^2\left(\partial_x^{\alpha-1}\frac{1}{r}\right)^2 + 2\alpha x\partial_x^{\alpha-1}\frac{1}{r}\partial_x^\alpha\frac{1}{r} + x^2\left(\partial_x^\alpha\frac{1}{r}\right)^2 \\
\left(\partial_y^\alpha \frac{y}{r}\right)^2 &= \alpha^2\left(\partial_y^{\alpha-1}\frac{1}{r}\right)^2 + 2\alpha y\partial_y^{\alpha-1}\frac{1}{r}\partial_y^\alpha\frac{1}{r} + y^2\left(\partial_y^\alpha\frac{1}{r}\right)^2 \\
\left(\partial_z^\alpha \frac{z}{r}\right)^2 &= \alpha^2\left(\partial_z^{\alpha-1}\frac{1}{r}\right)^2 + 2\alpha z\partial_z^{\alpha-1}\frac{1}{r}\partial_z^\alpha\frac{1}{r} + z^2\left(\partial_z^\alpha\frac{1}{r}\right)^2
\end{aligned} \tag{17}$$

Replacing Eq. (17) with the explanatory form of the Legendre polynomials, for the following terms $\partial_x^{\alpha-1}\frac{1}{r}$ and $\partial_x^\alpha\frac{1}{r}$.

$$\begin{aligned}
\partial_x^\alpha\frac{1}{r} &= (-1)^\alpha\alpha!\left(x^2 + y^2 + z^2\right)^{-\frac{(\alpha+1)}{2}} P_\alpha\left(\frac{x}{\left(x^2 + y^2 + z^2\right)^{1/2}}\right) \\
\partial_x^{\alpha-1}\frac{1}{r} &= (-1)^{\alpha-1}(\alpha-1)!\left(x^2 + y^2 + z^2\right)^{-\frac{\alpha}{2}} P_{\alpha-1}\left(\frac{x}{\left(x^2 + y^2 + z^2\right)^{1/2}}\right)
\end{aligned} \tag{18}$$

Also, knowing that for each $\alpha \geq 0$, the maximum value of $P_\alpha(1) = 1$. We can write the following inequality

$$\begin{aligned}
x^2\left(\partial_x^\alpha\left(\frac{1}{r}\right)\right)^2 &\leq x^2(\alpha!)^2 r^{-2(\alpha+1)} \\
2\alpha x\partial_x^{\alpha-1}\frac{1}{r}\partial_x^\alpha\frac{1}{r} &\leq 2x\alpha(\alpha!)(\alpha-1)!(-1)^{2\alpha-1}r^{-2\alpha-1} \\
\alpha^2\left(\partial_x^{\alpha-1}\left(\frac{1}{r}\right)\right)^2 &\leq \alpha^2((\alpha-1)!)^2 r^{-2\alpha}
\end{aligned} \tag{19}$$

Grouping terms for $\left(\partial_x^\alpha\frac{x}{r}\right)^2$, $\left(\partial_y^\alpha\frac{y}{r}\right)^2$ and $\left(\partial_z^\alpha\frac{z}{r}\right)^2$ we have the next expressions.

$$\begin{aligned}
\left(\partial_x^\alpha \frac{x}{r}\right)^2 &\leq r^{-2\alpha}\left[\frac{x^2(\alpha!)^2}{r^2} - \frac{2x(\alpha!)^2}{r} + \alpha^2((\alpha-1)!)^2\right] \\
\left(\partial_y^\alpha \frac{y}{r}\right)^2 &\leq r^{-2\alpha}\left[\frac{y^2(\alpha!)^2}{r^2} - \frac{2y(\alpha!)^2}{r} + \alpha^2((\alpha-1)!)^2\right] \\
\left(\partial_z^\alpha \frac{z}{r}\right)^2 &\leq r^{-2\alpha}\left[\frac{z^2(\alpha!)^2}{r^2} - \frac{2z(\alpha!)^2}{r} + \alpha^2((\alpha-1)!)^2\right]
\end{aligned} \tag{20}$$

The module of $\left|\partial_{\mathbf{x}}^{\alpha}\mathbf{u}^0\right|$ is given by $\left|\partial_{\mathbf{x}}^{\alpha}\mathbf{u}^0\right| = \left(\left(\partial_x^{\alpha}\frac{x}{r}\right)^2 + \left(\partial_y^{\alpha}\frac{y}{r}\right)^2 + \left(\partial_z^{\alpha}\frac{z}{r}\right)^2\right)^{1/2}$. Simplifying and placing the terms of Eq. (20) we have

$$\left|\partial_{\mathbf{x}}^{\alpha}\mathbf{u}^0\right| \le r^{-2\alpha}\left[3(\alpha!)^2 + \alpha^2((\alpha-1)!)^2 - \frac{2(x+y+z)(\alpha!)^2}{r}\right]$$

Taking into consideration that $\left|\frac{x}{r}\right| \le 1,\ \left|\frac{y}{r}\right| \le 1,\ \left|\frac{z}{r}\right| \le 1$ the last term $\left|\partial_{\mathbf{x}}^{\alpha}\mathbf{u}^0\right|$ can be easily written that.

$$\left|\partial_{\mathbf{x}}^{\alpha}\mathbf{u}^0\right| \le \frac{2(\alpha!)^2}{r^{2\alpha}}\left[2 + \left|\frac{x}{r}\right| + \left|\frac{y}{r}\right| + \left|\frac{z}{r}\right|\right]$$
$$\left|\partial_{\mathbf{x}}^{\alpha}\mathbf{u}^0\right| \le \frac{10(\alpha!)^2}{r^{2\alpha}}$$

It is verified that there exists $C_\alpha = 10(\alpha!)^2$ such that if $r \to 0$, then $\left|\partial_{\mathbf{x}}^{\alpha}\mathbf{u}^0\right| \to 0$. Thus, we proved requirement (4). ■

According to Mathematics, and giving an integral physical structure to the study, we need to prove that there are the spatial and temporal derivatives of the velocity and pressure components, satisfying the requirement (5).

Proposition 5 *Requirement (5). The velocity can be obtained from:* $\mathbf{u}(x,y,z,t) = -2\nu\frac{\nabla P}{P}$ *and each of the components* u_x, u_y *and* u_z *are infinitely derivable.*

$$\mathbf{u}(x,y,z,t) = 2\nu^2\left(\frac{x}{r^2}, \frac{y}{r^2}, \frac{z}{r^2}\right) \qquad \left((x,y,z)\in\mathbb{R}^3\right)$$
$$P(x,y,z,t) = \frac{1}{1+e^{\frac{p_0}{2\eta}t-\mu r}} = \frac{2}{\mu r} \tag{21}$$

Proof. Taking partial derivatives for $\partial_x^n\left(\frac{x}{r^2}\right)$, $\partial_y^n\left(\frac{x}{r^2}\right)$ and $\partial_z^n\left(\frac{x}{r^2}\right)$.

$$\partial_x^n\left(\frac{x}{r^2}\right) = n\partial_x^{n-1}\left(\frac{1}{r^2}\right) + x\partial_x^n\left(\frac{1}{r^2}\right)$$
$$\partial_y^n\left(\frac{x}{r^2}\right) = n\partial_y^{n-1}\left(\frac{1}{r^2}\right) + y\partial_y^n\left(\frac{1}{r^2}\right) \tag{22}$$
$$\partial_z^n\left(\frac{x}{r^2}\right) = n\partial_z^{n-1}\left(\frac{1}{r^2}\right) + z\partial_z^n\left(\frac{1}{r^2}\right)$$

Recalling the derivatives of special functions, it is verified that the derivative C^∞ exists. These derivatives appear as a function of the Legendre polynomials $P_n(.)$.

$$\partial_x^n\left(\frac{1}{r^2}\right) = (-1)^n n!\left(x^2+y^2+z^2\right)^{-(n+1)}P_n\left(\frac{x}{(x^2+y^2+z^2)}\right)$$
$$\partial_y^n\left(\frac{1}{r^2}\right) = (-1)^n n!\left(x^2+y^2+z^2\right)^{-(n+1)}P_n\left(\frac{y}{(x^2+y^2+z^2)}\right) \tag{23}$$
$$\partial_z^n\left(\frac{1}{r^2}\right) = (-1)^n n!\left(x^2+y^2+z^2\right)^{-(n+1)}P_n\left(\frac{z}{(x^2+y^2+z^2)}\right)$$

There are the spatial derivatives n and the time derivative which is similar to Eq. (25). ■

Proposition 6 *Requirement (5). The pressure is totally defined by the equivalence* $p(x,y,z,t) = p_0 P(x,y,z,t)$ *and is infinitely differentiable in each of its components.*

$$p(x,y,z,t) = p_0 P(x,y,z,t) \qquad \left((x,y,z) \in \mathbb{R}^3\right) \tag{24}$$

Proof. Taking partial derivatives for $\partial_x^n\left(\frac{1}{r}\right)$, $\partial_y^n\left(\frac{1}{r}\right)$ and $\partial_z^n\left(\frac{1}{r}\right)$, recalling the derivatives of special functions of Eq. (16), it is shown that the derivative C^∞. We only have to find the time derivatives: $\partial_t^n(p_0 P) = p_0 \partial_t^n(P)$. Using Eq. (21) for P, we have.

$$\begin{aligned}
\partial_t P &= (-k)P(1-P) \\
\partial_t^2 P &= (-k)^2(1-2P)P(1-P) \\
\partial_t^3 P &= (-k)^3\left(1-6P+6P^2\right)P(1-P) \\
\partial_t^4 P &= (-k)^4\left(1-14P+36P^2-24P^3\right)P(1-P) \\
\partial_t^5 P &= (-k)^5\left(1-30P+150P^2-240P^3+120P^4\right)P(1-P) \\
\partial_t^n(P) &= \partial_t\left(\partial_t^{n-1}(P)\right)
\end{aligned} \tag{25}$$

It is always possible to find the derivative $\partial_t^n(P)$ as a function of the previous derivative, since the resulting polynomial of each derivative $n-1$ is of degree n. ■

Proposition 7 *Requirement (6). The energy must be limited in a defined volume and fundamentally it must converge at any time, such that* $t \geq 0$.

$$\int_{\mathbb{R}^3} |\mathbf{u}(x,y,z,t)|^2 dxdydz \leq C \quad \textit{for all}\ t \geq 0 \qquad (\textit{bounded energy}).$$

Proof. We will use the explicit form of velocity given in Eq. (21) $\mathbf{u}(x,y,z,t) = 2\nu\mu(1-P)\nabla r$, to obtain the vector module: $|\mathbf{u}|^2 = 4\nu^2\mu^2(1-P)^2$. Rewriting Eq. (21), and applying a change of variable in: $dxdydx = 4\pi r^2 dr$.

$$\int_{\mathbb{R}^3} |\mathbf{u}(x,y,z,t)|^2 dxdydz = 16\pi\nu^2\mu^2 \int_{r_0}^{\infty} r^2(1-P)^2 dr \tag{26}$$

Making another change of variable $dP = \mu P(1-P)dr$. Using (10), replacing $r^2 = \left(\frac{2}{\mu P}\right)^2$ we have

$$\begin{aligned}
\int_{\mathbb{R}^3} |\mathbf{u}(x,y,z,t)|^2 dxdydz &= 16\pi\nu^2\mu^2 \int_{P_0}^{P\infty} \left(\frac{2}{\mu P}\right)^2 (1-P)^2 \frac{dP}{\mu P(1-P)} \\
&= \frac{64\pi\nu^2}{\mu} \int_{P_0}^{P\infty} \frac{1-P}{P^3} dP
\end{aligned} \tag{27}$$

Where radius $r \to \infty$, when $t \geq 0$, we have $\lim_{r\to\infty} P = \lim_{r\to\infty} \frac{1}{1+\frac{exp(kt)}{\exp(\mu r)}} = P_\infty = 1$. Moreover, physically if $r \to r_0 \approx 0$ then $t \to 0$ we have $\lim_{r\to 0} P = \lim_{r\to 0} \frac{1}{1+\frac{exp(kt)}{\exp(\mu r)}} = P_0 = \frac{1}{2}$. Here, a probability $\frac{1}{2}$ represents maximum entropy.

$$\begin{aligned}
\int_{\mathbb{R}^3} |\mathbf{u}(x,y,z,t)|^2 dxdydz &= \frac{64\pi\nu^2}{\mu} \int_{1/2}^{1} \frac{1-P}{P^3} dP = \frac{64\pi\nu^2}{\mu} \left[\frac{2P-1}{2P^2}\right]_{1/2}^{1} \\
\int_{\mathbb{R}^3} |\mathbf{u}|^2 dxdydz &\leq \frac{32\pi\nu^2}{\mu} \qquad \text{forall } t \geq 0
\end{aligned} \tag{28}$$

In this way the value of the constant C is $C = \frac{32\pi\nu^2}{\mu}$. Verifying the proposition (6) completely. In general, Eq. (10) can be written $f(t, r + r_0) = \frac{1}{1+e^{kt-\mu(r+r_0)}} - \frac{2}{\mu\mu(r+r_0)} = 0$ and in this way discontinuities are avoided when $r \to 0$, but this problem does not occur since in the atomic nucleus $r_0 < r < 1.2A^{1/3}$ is satisfied. ■

Lemma 8 *The irrotational field represented by the logistic probability function $P(x, y, z, t)$ associated with the velocity $\mathbf{u} = -2\nu \frac{\nabla P}{P}$, can produce vortices, due to the stochastic behavior of the physical variables p_0, η, μ. These stochastic variations are in orders lower than the minimum experimental value.*

Proof. The implicit function representing the solution of the Navier–Stokes 3D equation, $\frac{1}{1+e^{-\mu\left(r-\frac{k}{\mu}t\right)}} = \frac{2}{\mu r}$ depends on the values of initial pressure p_0, viscosity ν and attenuation coefficient μ. Due to Heisenberg uncertainty principle, these parameters have a variation when we measure and use them, as is the case of the estimate of $\xi = r - \phi(t, k, \mu)$. The function $\phi(t, k, \mu) = \frac{k}{\mu}t$, expressly incorporates these results, when $-\infty < \xi < +\infty$. The physical and mathematical realities are mutually conditioned and allow for these surprising results. For a definite t there exist infinities (x, y, z) that hold the relationship $r = (x^2 + y^2 + z^2)^{1/2}$. Moreover, for a definite r there are infinities t that respect the fixed-point theorem and create spherical trajectories. When the physical variables k, μ vary, even at levels of 1/100 or 1/1000, they remain below the minimum variation of the experimental value. We could try to avoid the existence of trajectories on the spherical surface, for which we must assume that the fluid is at rest or it is stationary, which contradicts the Navier–Stokes 3D equation, where all fluid is in accelerated motion $\frac{\partial \mathbf{u}}{\partial t} \neq 0$. In short, if there are trajectories in the sphere as long as it is probabilistically possible, this is reduced to showing that the expected value of the radius $E[r|r \geq 0]$ exists and it is finite.

Derivation of $E(r|r \geq 0)$.

The logistic density function for ξ when $E(\xi) = 0$ and $Var(\xi) = \sigma^2$ is defined by. $h(\xi) = \frac{\mu \exp(-\xi)}{[1+\exp(-\xi)]^2}$, where $\frac{1}{\mu} = \sigma\sqrt{3}/\pi$ is a scale parameter. Given that $r = \phi(t, p_0, \eta, \mu) + \xi$ function for r is then $f(r) = \frac{\exp[-r-\phi(\bullet)/\tau]}{\tau[1+\exp(-(r-\phi(\bullet))/\tau]^2}$, to facilitate the calculations we put $\phi(\bullet) = \phi(t, k, \mu) = \frac{k}{\mu}t$. By definition, the truncated density for r when $r \geq 0$ is given by $f(r|r \geq 0) = \frac{f(r)}{P(r \geq 0)}$ for $r \geq 0$. Given that the cumulative distribution function for r is given by $F(r) = \frac{1}{1+\exp(kt-\mu r)}$, it follows that $P(r \geq 0) = 1 - F(0) = \frac{\exp(\phi(\bullet)))}{1+\exp(\phi(\bullet)))} = \frac{1}{1+\exp(-\phi(\bullet)))}$. The derivation of $E(r|r \geq 0)$ then proceeds as follows:

$$E(r|r \geq 0) = \int_0^\infty \mu r f(r|r \geq 0) dr = \frac{1}{P(r \geq 0)} \int_0^\infty \mu r \frac{\exp[kt - \mu r]}{\{1 + \exp[kt - \mu r]\}^2} dr$$
$$E(r|r \geq 0) = \frac{1}{P(r \geq 0)} \int_{1/2}^1 \frac{2}{\mu P} (\mu P(1-P)) \frac{dP}{P(1-P)} \tag{29}$$

We replaced in Eq. (29) $dP = \mu P(1-P)dr$ and $r^2 = \left(\frac{2}{\mu P}\right)^2$ of this manner we obtain

$$E(r|r \geq 0) = \frac{1}{P(r \geq 0)} \int_{1/2}^1 \frac{2}{\mu P} (\mu P(1-P)) \frac{dP}{P(1-P)} = \frac{1}{P(r \geq 0)} \frac{2}{\mu} \log(2) \tag{30}$$

where we have used the fact that

$$P(r \geq 0) = \frac{\exp kt}{1 + \exp kt}$$

$$E(r|r \geq 0) = \frac{1}{P(r \geq 0)} \frac{2}{\mu} \log(2) \leq \frac{2}{\mu} \log(2) \quad \text{for all } t \geq 0 \tag{31}$$

where the last equality follows from an application of the L'Hopital's rule $P(r \geq 0) = \lim_{t \to \infty} \frac{\exp kt}{1+\exp kt} = 1$. ■

3. Results

The main results of applying the Navier Stokes equation to the atomic nucleus, which behaves like an incompressible nuclear fluid, are:

- The nuclear force and the Navier Stokes force are related.
- The Cross Sections in Low energy X ray can explain the Golden Ratio $\left(\frac{\sigma_1}{\sigma_2}\right)$ that appears in the femtoscope.
- Navier Stokes Equation and Cross Section in Nuclear Physics.
- The principle of the femtoscope explains that low energy X-rays produce resonance in layer K.

3.1 The nuclear force and the Navier Stokes force are related

Firstly, we will use the concepts of Classic Mechanics and the formulation of the Yukawa potential, $\Phi(r) = \frac{g}{4\pi r}(A-1)e^{-\mu r}$ to find the nuclear force exerted on each nucleon at interior of the atomic core $\mathbf{F}_N = -\nabla\Phi(r)$. Also, replace the terms of the potential $e^{-\mu r} = \frac{1-P}{P}$ and $\frac{1}{r} = \frac{\mu}{2}P$ by the respective terms already obtained in Eq. (10).

$$\Phi(r) = \frac{g(A-1)}{4\pi r} e^{-\mu r} = \frac{g\mu(A-1)}{8\pi}(1-P) \tag{32}$$

The general form of the Eq. (32), is a function of (x, y, z, t).

$$\Phi(r,t) = \frac{g(A-1)}{4\pi r} e^{kt-\mu r} = \frac{g\mu(A-1)}{8\pi}(1-P(r,t)) \tag{33}$$

Secondly, we will obtain the Navier Stokes force equation given by:

$$\frac{d\mathbf{u}}{dt} = \frac{\partial \mathbf{u}}{\partial t} + (\mathbf{u}.\nabla)\mathbf{u} = \frac{\partial \mathbf{u}}{\partial t} + 2\nu^2 \nabla\left(\frac{|\nabla P|^2}{P^2}\right) = -2\mu\nu k P(1-P)\nabla \mathbf{r} \tag{34}$$

Theorem 9 *The Nuclear Force and Navier Stokes Force are proportional inside the atomic nucleus* $\mathbf{F}_N = C\mathbf{F}_{NS}$.

Proof. Eq. (34) rigorously demonstrated by theorems and propositions 1 through 8, represent the acceleration of a particle within the atomic nucleus. According to Classical Mechanics the force of Navier Stokes applied to a particle of mass m, would have the form:

$$\mathbf{F}_{NS} = m\frac{d\mathbf{u}}{dt} = -2m\mu\nu kP(1-P)\nabla\mathbf{r}. \tag{35}$$

∎

Proof. The nuclear force on its part would be calculated as follows $\mathbf{F}_N = -\nabla\Phi(r)$.

$$\mathbf{F}_N = -\nabla\Phi(r) = -\frac{g\mu}{8\pi}(A-1)\nabla P. \tag{36}$$

Replacing the term $\nabla P = \mu(P - P^2)\nabla r$ of Eq. (7), we obtain

$$\mathbf{F}_N = -\nabla\Phi(r) = -\frac{g\mu^2}{8\pi}(A-1)P(1-P)\nabla\mathbf{r}. \tag{37}$$

It is possible to write nuclear force as a function of speed.

$$\mathbf{F}_N = -\nabla\Phi(r) = -\frac{g\mu^2}{8\pi}(A-1)P(1-P)\nabla\mathbf{r}.$$

Finally, we can show that the nuclear force and force of Navier Stokes differ at most in a constant C. Equating (35) and (37), we find the value g as a function of the parameters nuclear viscosity ν, attenuation μ and growth coefficient of the nuclear reaction k, nucleon mass m and $C \neq 1$.

$$g = \frac{16m\pi\nu k}{\mu(A-1)}C \tag{38}$$

∎

3.2 Cross section and Golden ratio $\left(\frac{\sigma_1}{\sigma_2}\right)$ are important elements of the femtoscope

According to NIST and GEANT4 [17], current tabulations of $\frac{\mu}{\rho}$ rely heavily on theoretical values for the total cross section per atom, σ_{tot}, which is related to $\frac{\mu}{\rho}$ by the following equation:

$$\frac{\mu}{\rho} = \frac{\sigma_{tot}}{uA} \tag{39}$$

In (Eq. 39), $u(= 1.6605402 \times 10^{-24} gr)$ is the atomic mass unit (1/12 of the mass of an atom of the nuclide 12C)[4].

The attenuation coefficient, photon interaction cross sections and related quantities are functions of the photon energy. The total cross section can be written as the sum over contributions from the principal photon interactions

$$\sigma_{tot} = \sigma_{pe} + \sigma_{coh} + \sigma_{incoh} + \sigma_{trip} + \sigma_{ph.n} \tag{40}$$

Where σ_{pe} is the atomic photo effect cross section, σ_{coh} and σ_{incoh} are the coherent (Rayleigh) and the incoherent (Compton) scattering cross sections, respectively, σ_{pair} and σ_{trip} are the cross sections for electron-positron production in the fields of the nucleus and of the atomic electrons, respectively, and $\sigma_{ph.n}$. is the photonuclear cross section 3,4.

We use data of NIST and simulations with GEANT4 for elements $Z = 11$ to $Z = 92$ and photon energies 1.0721×10^{-3} MeV to 1.16×10^{-1} MeV, and have been calculated according to:

$$\frac{\mu}{\rho} = \left(\sigma_{pe} + \sigma_{coh} + \sigma_{incoh} + \sigma_{trip} + \sigma_{ph.n}\right)/\mu A \tag{41}$$

The attenuation coefficient μ of a low energy electron beam $[10, 100]eV$ will essentially have the elastic and inelastic components. It despises Bremsstrahlung emission and Positron annihilation.

$$\sigma_{tot} = \sigma_{coh} + \sigma_{incoh} \tag{42}$$

3.3 The principle of the femtoscope explains that low energy X-rays produce resonance in K-shell

A resonance region is created in a natural way at the K-shell between the nucleus and the electrons at S-level. The condition for the photons to enter in the resonance region is given by $r_a \geq r_n + \lambda$. This resonance region gives us a new way to understand the photoelectric effect. There is experimental evidence of the existence of resonance at K-level due to photoelectric effect, represented by the resonance cross section provided by NIST and calculated with GEANT4 for each atom. In the present work we focus on the resonance effects but not on the origin of resonance region.

The resonance cross section is responsible for large and/or abnormal variations in the absorbed radiation $(I_2 - I_1)$.

$$\frac{I_2 - I_1}{\frac{I + I_1}{2}} = -\frac{\rho r}{\mu A}(\sigma_2 - \sigma_1) \tag{43}$$

Theorem 10 *Resonance region. The resonance cross section is produced by interference between the atomic nucleus and the incoming X-rays inside the resonance region, where the boundaries are the surface of the atomic nucleus and K-shell.*

The cross section of the atomic nucleus is given by:

$$\sigma_{r_n} = 4\pi r_n^2 = 4\pi A^{2/3} r_n^2 \tag{44}$$

The photon cross section at K-shell depends on the wave length and the shape of the atomic nucleus:

$$\sigma_{r_n+\lambda} = 4\pi(r_n + \lambda)^2 \tag{45}$$

Subtracting the cross sections (13) and (14) we have:

$$\sigma_\lambda = \sigma_{r_n+\lambda} - \sigma_{r_n} = 4\pi\left(2r_n\lambda + \lambda^2\right) = 4\pi\left(2r_p\lambda + 2\left(r_n - r_p\right)\lambda + \lambda^2\right) \tag{46}$$

The resonance is produced by interactions between the X-rays, the K-shell electrons and the atomic nucleus. The cross sections corresponding to the nucleus is weighted by probability pn and should have a simple dependence of an interference term. This last depends on the proton radius rp or the difference between the nucleus and proton radius $(r_n - r_p)$ according to the following relation

$$Max(\sigma_2 - \sigma_1) = \left(\frac{\sigma_1}{\sigma_2}\right)^{-b}(\sigma_2 - \sigma_1) = 4\pi\left(2r_p\lambda\right) \tag{47}$$

We note that left hand side of Eqs. (46) and (47) should have a factor larger than one due to resonance. The unique factor that holds this requirement is $\left(\frac{\sigma_1}{\sigma_2}\right)^b$ Where a, b constants.

$$\frac{8\pi\bar{r}\lambda}{(\sigma_2 - \sigma_1)} = a\left(\frac{\sigma_1}{\sigma_2}\right)^b \tag{48}$$

After performing some simulations it is shown that the thermal a represents the dimensionless Rydberg constant $a = R_\infty = 1.0973731568539 * 10^7$.

$$\frac{8000\pi\bar{r}\lambda}{(\sigma_2 - \sigma_1)} = R_\infty\left(\frac{\sigma_1}{\sigma_2}\right)^{2.5031} \tag{49}$$

We use $\sigma_1(Z), \sigma_2(Z)$ for represent cross section in resonance, R_∞ is the generalized Rydberg constant for all elements of periodic table, and Z atomic number.

Last equation with a $R^2 = 0.9935$ was demonstrated and constructed using the elements of the solution of the Navier Stokes equations.

$$\left(\frac{\sigma_1}{\sigma_2}\right) = 0.0021Z + 0.0696$$

This equation was obtained with a $R^2 = 0.9939$, and indicates that the ratio of the effective sections fully explain each element of the periodic table.

$$E^* = 2 * 10^{-5}Z^2 - 0.0003Z + 0.004$$

This equation obtained for a $R^2 = 0.9996$, complements the system of equations that allow to know the simulation values as a function of Z for $\sigma_1(Z), \sigma_2(Z)$ and $E^*(Z)$, where $\sigma_1(Z) < \sigma_2(Z)$. The Femtoscope equations further demonstrate that energy ($E^* = \min E$) is minimum and Shannon entropy ($S^* = \max S$) is maximum in resonance, because in equilibrium $\sigma_1(Z_i) = \sigma_2(Z_i)$.

The radius of the neutron can be obtained using Eq. (49) in the following way.

$$\bar{r} = \frac{R_\infty}{8000\pi a}$$

$$r_N = \frac{A}{N}\bar{r} + \frac{Z}{N}r_p$$

3.4 Navier Stokes equation and cross section in nuclear physics

The speed needs to be defined as $\mathbf{u} = -2\nu\frac{\nabla P}{P}$, where $P(x, y, z, t)$ is the logistic probability function $P(x, y, x, t) = \frac{1}{1+e^{kt-\mu r}}$, $r = (x^2 + y^2 + z^2)^{1/2}$ defined in $\left((x, y, z) \in \mathbb{R}^3, t \geq 0\right)$ This P is the general solution of the Navier Stokes 3D equations, which satisfies the conditions (50) and (51), allowing to analyze the dynamics of an incompressible fluid.

$$\frac{\partial \mathbf{u}}{\partial t} + (\mathbf{u}.\nabla)\mathbf{u} = -\frac{\nabla p}{\rho_0} \qquad \left((x, y, z) \in \mathbb{R}^3, t \geq 0\right) \tag{50}$$

Where, $\mathbf{u} \in \mathbb{R}^3$ an known velocity vector, ρ_0 constant density of fluid and pressure $p = p_0 P \in \mathbb{R}$.

With speed and pressure dependent on r and t. We will write the condition of incompressibility as follows.

$$\nabla.\mathbf{u} = 0 \qquad \left((x,y,z)\in\mathbb{R}^3, t\geq 0\right) \tag{51}$$

Theorem 11 *The velocity of the fluid given by:* $\mathbf{u} = -2\nu\frac{\nabla P}{P}$, *where* $P(x,y,z,t)$ *is6the logistic probability function* $P(x,y,x,t) = \frac{1}{1+e^{kt-\mu\left(x^2+y^2+z^2\right)^{1/2}}}$, *defined in* $\left((x,y,z)\in\mathbb{R}^3, t\geq 0\right)$ *is the general solution of the Navier Stokes equations, which satisfies conditions (50) and (51).*

Proof. Firstly, we will make the equivalence $\mathbf{u} = \nabla\theta$ and replace it in Eq. (50). Taking into account that $\nabla\theta$ is irrotational, $\nabla \times \nabla\theta = 0$, we have.

$$(\mathbf{u}.\nabla)\mathbf{u} = (\nabla\theta.\nabla)\nabla\theta = \frac{1}{2}\nabla(\nabla\theta.\nabla\theta) - \nabla\theta \times (\nabla \times \nabla\theta) = \frac{1}{2}\nabla(\nabla\theta.\nabla\theta),$$

We can write,

$$\nabla\left(\frac{\partial\theta}{\partial t} + \frac{1}{2}(\nabla\theta.\nabla\theta)\right) = \nabla(-p)$$

It is equivalent to,

$$\frac{\partial\theta}{\partial t} + \frac{1}{2}(\nabla\theta.\nabla\theta) = -\frac{\Delta p}{\rho_0}$$

where Δp is the difference between the actual pressure p and certain reference pressure p_0. Now, replacing $\theta = -2\nu\ln(P)$, Navier Stokes equation becomes.

$$\frac{\partial P}{\partial t} = \frac{\Delta p}{\rho_0}P \tag{52}$$

The external force is zero, so that there is only a constant force F due to the variation of the pressure on a cross section σ. Where σ is the total cross section of all events that occurs in the nuclear surface including: scattering, absorption, or transformation to another species.

$$\begin{aligned} F &= p\sigma_2 = p_0\sigma_1 \\ \Delta p &= p - p_0 = \left(\frac{\sigma_1}{\sigma_2} - 1\right)p_0 = -(1-P)p_0 \end{aligned} \tag{53}$$

putting (52) in (53) we have

$$\frac{\partial P}{\partial t} = -\mu k(1-P)P \tag{54}$$

In order to verify Eq. (51), $\nabla.\mathbf{u} = 0$, we need to obtain $\nabla r = \left(\frac{x}{r}, \frac{y}{r}, \frac{z}{r}\right)$, $\nabla^2 r = \nabla.\nabla r = \frac{\left(y^2+z^2\right)+\left(x^2+z^2\right)+\left(x^2+y^2\right)}{\left(x^2+y^2+z^2\right)^{3/2}} = \frac{2}{r}$.

$$\nabla.u = -2\nu\nabla.\frac{\nabla P}{P} = -2\nu\mu\nabla((1-P)\nabla r) \tag{55}$$

Replacing the respective values for the terms: $\nabla^2 P$ and $|\nabla P|^2$ of Eq. (55). The Laplacian of P can be written as follows.

$$\begin{aligned}\nabla^2 P &= \mu(1-2P)\nabla P.\nabla r + \mu\left(P - P^2\right)\nabla^2 r \\ &= \mu^2(1-2P)\left(P-P^2\right)|\nabla r|^2 + \mu\left(P-P^2\right)\nabla^2 r \\ &= \mu^2(1-2P)\left(P-P^2\right) + \mu\left(P-P^2\right)\frac{2}{r}\end{aligned} \tag{56}$$

Using gradient $\nabla P = \mu\left(P-P^2\right)\nabla r$, modulus $|\nabla P|^2 = \mu^2\left(P-P^2\right)^2|\nabla r|^2$ and $\nabla^2 P$ in (56).

$$\left[\frac{\nabla^2 P}{P} - \frac{|\nabla P|^2}{P^2}\right] = 0 \tag{57}$$

Replacing Eqs. (55) and (56) in (57) we obtain the main result of the Navier Stokes equations, the solution represents a fixed point of an implicit function $f(t,r)$ where $f(t,r) = P - \frac{2}{\mu r} = 0$.

$$P = \frac{1}{1+e^{kt-\mu\left(x^2+y^2+z^2\right)^{1/2}}} = \frac{2}{\mu\left(x^2+y^2+z^2\right)^{1/2}} \qquad \left((x,y,z)\in\mathbb{R}^3, t\geq 0\right) \tag{58}$$

■

An important result of the Navier Stokes 3D equation, applied to the nuclear fluid of an atom, allows us to advance our understanding of nuclear dynamics and nuclear force.

Corollary 12 *The nuclear decay constant k is determined by the nuclear pressure p_0, and the dynamic nuclear viscosity η, as follows: $k = \frac{p_0}{2\eta}$.*

Proof. A nuclear decay is a reaction of degree 1, which is explained by the exponential law $N(t) = N_0 e^{-kt}$. Rewriting Eq. (58) as a function of the initial nuclear pressure p_0 and the nuclear viscosity η we can prove that they are related to the nuclear decay constant, in an intrinsic way and allow to explain dynamic nuclear phenomena.

$$P(x,y,z,t) = \frac{1}{1+e^{\frac{p_0}{2\eta}t-\mu\left(x^2+y^2+z^2\right)^{1/2}}} = \frac{2}{\mu\left(x^2+y^2+z^2\right)^{1/2}} \qquad \left((x,y,z)\in\mathbb{R}^3, t\geq 0\right) \tag{59}$$

■

3.5 Model of nuclear pressure

The work in the nuclear fluid is given by the variation of the energy necessary to form that atomic nucleus, that is, by the excess mass required in the process. Using Eq. (59) we can find the explicit form of nuclear pressure $p = p_0 P = p_0 \frac{2}{\mu r}$.

$$\begin{aligned} W &= \Delta E_{mass-excess}(Z,A) = \int_0^{r_0}\int_0^{\pi}\int_0^{2\pi} p dV = \int_0^{r_0}\int_0^{\pi}\int_0^{2\pi} p_0 \frac{2}{\mu r}\left(r^2 \cos\theta dr d\theta d\phi\right) \\ W &= \Delta E_{mass-excess}(Z,A) = p_0 \frac{2}{\mu}\int_0^{r_0} r dr \int_0^{\pi} \sin\theta d\theta \int_0^{2\pi} d\phi = 4\pi p_0 \frac{r_0^2}{\mu} \end{aligned} \tag{60}$$

Applying the mean value theorem of integrals, we know that there is a mean value of the nuclear pressure $\overline{p}$ and the volume of the atomic nucleus $\overline{V}$ of the integral (60)

$$W = \Delta E_{mass-excess}(Z, A) = \int_0^{r_0}\int_0^{\pi}\int_0^{2\pi} pdV = p(\zeta)V(r_0) \tag{61}$$

that according to Quantum Mechanics, $p(\zeta), V(r_0)$ are the observable values. By this restriction we can equalize Eq. (60), (61) and find the value of the initial nuclear pressure as a function of the experimentally measured nuclear pressure: $p(\zeta)\frac{4\pi}{3}r_0^3 = 4\pi p_0 \frac{r_0^2}{\mu}$ with $r_0 = 1.2A^{1/2}fm$, from where:

$$p(\zeta) = \frac{\Delta E_{mass-excess}(Z, A)}{\frac{4}{3}\pi r_0^3} = \frac{3p_0}{\mu r_0}. \tag{62}$$

For Yukawa's potential, it is often assumed $\mu \approx \frac{1}{r_0}$.

3.6 Model of nuclear viscosity

Using the fundamental expression $k = \frac{p_0}{2\eta} = \frac{1}{T_{1/2}}$ obtained from the resolution of the Navier Stokes 3D equations, Eqs. (58) and (59), it is possible to find the average or most probable values of the variables involved pressure, nuclear viscosity and nuclear decay constant as follows: $\overline{k} = \frac{\overline{p}}{2\overline{\eta}}$.

Explicitly we find the nuclear viscosity as a function of the nuclear decay constant $(k = \overline{k})$ and the average value of the nuclear pressure $\overline{p}$ as follows:

$$\overline{\eta} = \frac{\overline{p}}{2\overline{k}} \tag{63}$$

There is another experimental way of determining nuclear viscosity, through the fuzziness of alpha particles, protons or neutrons ejected in a nuclear decay using the fluid velocity equation $\mathbf{u} = -2\nu\mu(1-P)\nabla r$, in which modulo $\mathbf{u} = -2\nu\mu(1-P)\nabla r$, replacing the dynamic viscosity $\nu = \frac{\eta}{\rho_0}\left[\frac{m^2}{s}\right]$, we obtain: $|\mathbf{u}| = 2\left(\frac{\eta}{\rho_0}\right)\mu(1-P)$.

So the second way to find the nuclear viscosity has the form:

$$\eta = \frac{|\mathbf{u}|\rho_0}{2\mu(1-P)} = \frac{|\mathbf{u}|\rho_0 rP}{4(1-P)} \tag{64}$$

Eq. (64) has a more complicated form and depends on the speed of the fluid particles and the radius from which these particles leave.

3.7 Calculation of nuclear pressure and viscosity

The nuclear decay constant k is determined by the nuclear pressure p_0, and the dynamic nuclear viscosity η, as follows: $k = \frac{p_0}{2\eta}$.

4. Discussion of results

4.1 Calculation of the pressure relation of proton and neutron

The fine-structure constant, α, has several physical interpretations, we use the most known.

The fine-structure constant, $\alpha = \frac{1}{137.035999174(35)}$ is the ratio of two energies: the energy needed to overcome the electrostatic repulsion between two protons a distance of $2r$ apart, and (ii) the energy of a single photon of wavelength $\lambda = 2\pi r$ (or of angular wavelength r; according to Planck relation).

$$\alpha' = \frac{e^2}{4\pi\varepsilon_0(2r)} / \frac{hc}{\lambda} = \frac{e^2}{4\pi\varepsilon_0(2r)} \frac{2\pi r}{hc} = \frac{e^2}{4\pi\varepsilon_0(2r)} \frac{r}{\hbar c} = \frac{1}{2} \frac{e^2}{4\pi\varepsilon_0 \hbar c} = \frac{1}{2}\alpha \tag{65}$$

We can find the relationship of energies between the proton and the atomic nucleus. Knowing that the two occupy the same nuclear volume V. This relationship is identical to $\frac{1}{\alpha'}$, since the atomic nucleus interacts with the proton through the electromagnetic field and the nuclear force.

$$\frac{p_p}{p_N} = \frac{p_p V}{p_N V} = \frac{2}{\alpha} = 274.07 \tag{66}$$

Thus, we already know the pressure relation between the proton and the atomic nucleus $\frac{p_p}{p_N} = \frac{2}{\alpha}$.

Now we find the relation of pressures between the neutron and the atomic nucleus, which are under the action of the same nuclear force, F.

$$\frac{p_n}{p_N} = \frac{F/\pi(0.84184)^2}{F/\pi\left(1.2A^{1/3}\right)^2} = \frac{\left(1.2A^{1/3}\right)^2}{(0.84184)^2} \tag{67}$$

For the chemical element with maximum nuclear pressure, ${}^{62}_{28}Ni$ we have $\frac{p_n}{p_N} = \frac{\left(1.2(62)^{1/3}\right)^2}{(0.84184)^2} = 31.830$.

If we divide Eq. (67) for Eq. (66) we have.

$$\frac{p_p}{p_n} = \frac{p_p}{p_N} / \frac{p_n}{p_N} = 274.07/31.830 = 8.6104$$

4.2 An action on the nuclear surface produces a reaction in the nuclear volume and vice versa

It is created by the friction between the layers of nucleons.

Theorem 13 *An action on the nuclear surface produces a reaction in the nuclear volume and vice versa.*

Proof. The volume and the nuclear surface are connected through the Gaussian divergence theorem and the Navier Stokes equations.

For an incompressible fluid, whose velocity field $\overleftarrow{u}(x, y, z)$ is given, $\overleftarrow{\nabla}.\overleftarrow{u} = 0$ is fulfilled.

Logically, the integral of this term remains zero, that is:

$$\iiint \overleftarrow{\nabla}.\overleftarrow{u} dxdydz = 0$$

Writing the Divergence theorem. $\iint \overleftarrow{u}.\overleftarrow{n} dS = \iiint \overleftarrow{\nabla}.\overleftarrow{u} dxdydz = 0$, the first term must be equal to zero, that is:

$$\iint \overleftarrow{u}.\overleftarrow{n}dS = \iint \left\|\overleftarrow{u}\right\| \left\|\overleftarrow{n}\right\| \cos(\alpha)dS = 0 \rightarrow \alpha = \frac{\pi}{2}$$

The only possible trajectory is circular, because in this case the vector $\overleftarrow{n}$ is perpendicular to the surface of the sphere. In this way the equation of the outer sphere corresponding to the surface is: $x^2 + y^2 + z^2 = 1.2A^{1/3}$.

Within the nuclear fluid there are layers of nucleons that move in spherical trajectories. ■

5. Conclusions

High energy physics is the guide to low energy physics, because, in certain processes such as in the measurement of the internal pressure of protons and neutrons. However, we demonstrate that there is a trend compatibility of the two in the characterization of the atomic nucleus.

The Femtomathematics corresponds to the tools and the logic that allows to calculate the parameters of the order of $10^{-15}m$. In an indirect way, it is in full correspondence with Femtophysics.

The nuclear viscosity at equilibrium is much larger than the nuclear viscosity at the moment of nuclear decay, which is totally logical, because at equilibrium the atomic nucleus is totally compact in pressure and density, while at the moment protons, neutrons, alpha particles come out of the atomic nucleus from a nuclear decay, indicating that the nuclear viscosity decreased.

Acknowledgements

1. To the participants of the congress: The 4th International Conference on Materials Sciences and Nanomaterials (ICMSN 2020), Cambridge, United Kingdom during July 08-10, 2020.

2. To the research directorate (DI) of the Central University of Ecuador, as well as the group of the

GIIP process of the Faculty Chemical Engineering, UCE.

Conflicts of interest

The author declare no conflict of interest.

Author details

Edward Henry Jimenez
Chemical Engineering Faculty, Universidad Central del Ecuador, Quito, Ecuador

*Address all correspondence to: ehjimenez@uce.edu.ec

References

[1] Burkert VD. Elouadrhiri, L. & Girod, F.X (2018). The pressure distribution inside the proton. Nature. 2018;**557**: 396-399. DOI: https://doi.org/10.1038/s41586-018-0060-z

[2] P. E. Shanahan and W. Detmold (2019). Pressure Distribution and Shear Forces inside the Proton. Phys. Rev. Lett. 122, 072003 – Published 22 February 2019

[3] Ozel F, Freire P. Masses, Radii, and Equation of State of Neutron Stars. Ann. Rev. Astron. Astrophys. 2016;**401-440** (2016):54

[4] Pagels, H (1966). Energy-Momentum Structure Form Factors of Particles", Phys. Rev. 144, 1250–1260 (1966).

[5] Ji, X. D (1997). Deeply virtual Compton scattering", Phys. Rev. D 55, 7114–7125 (1997).

[6] Teryaev, O. V (2016). Gravitational form factors and nucleon spin structure", Front. Phys. (Beijing) 11, 111207-111207-8 (2016).

[7] Yeverechyahu A. Nuclear viscosity and widths of giant resonances. Annals of Physics. Volume 95, Issue 1. November. 1975:1975

[8] Wong. Introductory to Nuclear Physics. WILEY-VCH Verlag Gmbl I & Co. Weinheim: KGaA; 2004. p. 2004

[9] Loveland W, Morrissey D, Seaborg GT. Modern Nuclear Chemistry. Wiley-Interscience. p. 57: 2005mnc..book.L. In: ISBN 978–0–471-11532-8. 2006

[10] Pohl R et al. The size of the proton. Nature. 2010;**2010**(466):213-216

[11] Jiménez, Recalde, Jimenez Chacon (2017). Extraction of the Proton and Electron Radii from Characteristic Atomic Lines and Entropy Principles. Entropy 2017.

[12] Fefferman (2017). EXISTENCE AND SMOOTHNESS OF THE NAVIER–STOKES EQUATION CHARLES L. FEFFERMAN, http://www.claymath.org/millennium-problems/navier%E2%80%93stokes-equation Accesed 01/05/2017.

[13] Kulish VV. On the Relationship between Fluid Velocity and de Broglie's Wave Function and the Implications to the Navier – Stokes Equation. International Journal of Fluid Mechanics Research. 2002;**29**(1):2002

[14] Leray J, l'espace S l m d'u l v e. Acta Math. J. 1934:63

[15] Caffarelli L, Kohn R, Nirenberg L. Partial regularity of suitable weak solutions of the Navier–Stokes equations. Comm. Pure & Appl. Math. 1982;**35**:771-831

[16] Kerson Huang (1987). Statistical Mechanics, 2nd Edition, John Willey & Sons, 1987.

[17] Geant 4. (2016). Geant4 User's Guide for Application Developers Version: geant4 10.3. Publication date 9 December 2016.

[18] Oganessian and Utyonkov. Superheavy nuclei from 48Ca-induced reactions. Nuclear Physics A. 2015;**944** (2015)

Section 4

Nuclear Materials for Fission and Fusion Systems

Chapter 4

Accident Tolerant Materials for LMFR

Viacheslav Sergeevich Okunev

Abstract

Several concepts of fast reactors with liquid metal coolant (LMFR) are being developed in the world. Lead-cooled reactors are most preferred for the safe nuclear power of the future. Projects of such reactors are being developed in Russia (BREST), the European Union (the latest development is ALFRED), and the USA (a series of STAR low-power reactor designs). The potential capabilities of fuel, coolant, and structural materials considered for use in the core to increase safety have not been exhausted. There are still unused reserves that can significantly increase the self-protection of the reactor. This chapter presents the results of the analysis of the use of new types of nuclear fuel: based on ceramics and beryllium and ceramics and uranium nanopowder. Studies are being conducted on the possibility of optimizing the composition of lead coolant without isotope separation. The possibilities of improving the safety of LMFR with a coolant based on lead extracted from thorium ores are being investigated. The possibility of using tungsten coatings of the cladding of fuel pins deposited using low-temperature plasma spraying is analyzed. The composition of materials was optimized in terms of improving reactor safety. The proposed innovations will significantly increase the self-protection of the reactor from the totality of severe accidents.

Keywords: liquid metal-cooled fast reactor, mixed oxide fuel, mixed nitride fuel, uranium nanopowder, self-protection, anticipated transients without scram, void reactivity effect

1. Introduction

Currently, the world is considering two basic concepts of medium- and high-power liquid metal-cooled fast reactor (LMFR): sodium cooled and lead cooled. Solving the problem of core optimization with restrictions characterizing the safe completion of emergency conditions accompanied by a failure of emergency protection, formally it is possible to eliminate severe accidents in a reactor with lead coolant and mononitride fuel. (The author used the DRAGON-M code [1], which makes it possible to optimize the geometric characteristics of the LMFR for given materials.) For the sodium-cooled LMFR, it is impossible to exclude severe accidents with core destruction.

It is time to optimize core materials. First of all, within the framework of the systematic approach, the multicriteria problem of choosing the optimal coolant for a power LMFR and ranking coolants according to the degree of preference was solved [2, 3]. Then, the properties of core materials (density, thermal conductivity, etc.) were included in the control parameters of the DRAGON-M. Then it remains only to select materials in a given range of properties.

2. Background

2.1 Background of the development of new technologies

The recent decades have radically changed the opinion of scientists about laws linking the probability of an accident and its consequences and about ways to prevent technological disasters and, in particular, ways to ensure the safety of nuclear power facilities. This is based on the following objective factors:

1. As industrial society and nuclear power developed in the early 1980s, it was revealed that the dependence of the probability of accidents (including at nuclear reactors and facilities) on damage or consequences (human casualties or the costs of measures to prevent them) is close to the distribution Gauss. The most probable accidents are characterized by small damage and the unlikely ones—by large damage, that is, technological disasters should occur extremely rarely, and nonserious accidents—often. Accumulated by the end of the twentieth century, catastrophe statistics on industrial facilities and transport have somewhat changed the usual idea: the probability of accidents depends weakly on damage; in any case, this dependence is closer to the exponential distribution. There is reason to believe that in the near future, the exponential distribution will be replaced by a uniform distribution: severe accidents and minor emergency situations will be equally probable.

2. Prior to the first serious accidents at nuclear power plants (NPPs), there was confidence that all emergency situations with unacceptable releases of radioactive substances into the environment could be prevented by using technical (engineering) safety systems. After serious accidents at NPP, safety criteria came to the fore. The "engineering" philosophy has been replaced by the philosophy of "inherent safety."

3. An increase in the number of potentially dangerous objects will increase the likelihood of accidents. By the middle of the twenty-first century, Russia is planning a transition to large-scale nuclear energy. If, for example, the number of power units increases by 10 times, then the probability of an accident (in a first approximation) will increase by 10 times. For this reason, when designing new-generation reactors and NPPs, a deterministic approach is preferred [4].

4. Nuclear energy is relatively young; emergency statistics are unrepresentative, which casts doubt on the appropriateness of using probabilistic analysis to justify the safety of reactors and opens up wide possibilities for using deterministic and (or) phenomenological approaches.

2.2 The history of the issue

After the Three Mile Island accident (1979), the work to eliminate severe accidents at NPPs was intensified in the USA and Europe. By the mid-1980s, concepts were proposed for the PIUS light-water reactor (ASEA-Atom, Sweden), high-temperature reactors KWU/Interction (Germany), GCRA (USA), DYONISOS (Switzerland), and MSGR liquid-salt reactor (China). The aim of the development was to eliminate the destruction of the core in severe accidents.

Among the technologically developed reactors, liquid metal-cooled fast reactors proved to be the most attractive, as far as possible to avoid severe accidents. In the early 1980s, in the USA research began on theoretical and experimental modeling of accident transients without scram (ATWS). By early April 1986 (a few days before the accident at Chernobyl NPP in the USSR), the work on experimental modeling of ATWS regimes at the EBR-II reactor (Argonne National Laboratory, Idaho, USA) was largely completed, demonstrating high capabilities in ensuring self-protection of low-power reactors with sodium cooling. The results of these unique studies are published in [5–10]. Zirconium-doped metal fuel has been proposed as a possible accident-tolerant fuel for fast sodium reactors in the USA [11]. It is a high-density and heat-conducting fuel. By the early 1990s, projects of modular sodium-cooled fast reactors PRISM and SAFR with such fuel were considered as promising in the USA [11, 12].

After the accident at the Chernobyl NPP (1986), Soviet scientists began to solve the problem of excluding severe accidents at reactors. It was clear that experimental modeling was limited in relation to the problem of safe reactors. First of all, this applies to the full-scale modeling of the ATWS. Such experiments are expensive and difficult to perform. They pose a potential danger to the public and the environment.

Sodium is characterized by high chemical activity. Its use requires an intermediate circulation circuit. Reactor used a three-circuit cooling systems. The melting point of metal fuel doped with zirconium is relatively low. Russia went on the way of the development of the heavy coolant (lead), much less active chemically, and more heat-resistant mixed mononitride fuel (MN). MN fuel is a reasonable compromise between heat-resistant mixed oxide (MOX) and high-density and heat-conducting metal fuel (including zirconium doped).

By 1989, the Soviet Union had developed a conceptual design of the lead-cooled BRS-1000 power fast reactor fueled by MN. By 1993, another conceptual design of the BRS-300 pilot reactor was proposed. It was the predecessor of the BREST-OD-300 project (JCS "NIKIET," Moscow, Russia), which has been developing to date [13–15]. The BREST uses a two-circuit cooling systems.

2.3 Specific hazards of fast reactors

In addition to the general hazard factors, fast reactors are organically characterized by reactivity accidents (the reactor is prompt supercritical) and the possibility of implementing a positive void reactivity effect (VRE) when using a liquid metal coolant. VRE is realized when the core or part of it is drained. Consider these two factors.

2.3.1 Risk of reactivity accidents

When entering reactivity $\rho > \beta$ (β is the effective delayed neutron fraction), the reactor is prompt supercritical. The change in reactor power or neutron density n over time t can be estimated using the point kinetics equation:

$$\frac{dn}{dt} \approx \frac{(\rho - \beta)nk}{l}, \tag{1}$$

where k is the effective multiplication factor, l is the average lifetime of prompt neutrons, $l = l_\infty P$, l_∞ is the average lifetime of prompt neutrons in an infinite medium, and P is the non- leakage probability.

Average lifetime of prompt neutrons in an infinite medium

$$l_\infty = \frac{\Lambda_a}{v} = \frac{1}{v\Sigma_a}, \tag{2}$$

where Λ_a is the average path length of a neutron until absorption.

In a fast reactor, the average neutron velocity v is several orders of magnitude higher than in a thermal reactors, although the macroscopic absorption cross section

Σ_a is two orders of magnitude lower. As a result, in fast reactors $l \sim 10^{-7}$–10^{-5} s, in thermal reactors $l \sim 10^{-3}$ s, other things being equal (ρ, β), the *dn*/*dt* in fast neutron reactors is higher, i.e., the power or density of neutrons increases faster. Obviously, the use of a coolant with a small absorption cross section also contributes to the self-protection of a fast reactor from reactivity accidents. In this case, the advantages of thorium lead (^{208}Pb) are obvious.

Another feature of fast reactors is the presence of ^{239}Pu in the fuel as the main fissile nuclide. The effective delayed neutron fraction for ^{235}U is equal 0.68%. The effective delayed neutron fraction for ^{239}Pu is equal 0.217%. As a result, when the same reactivity is introduced, the difference (ρ – β) is greater in a reactor using ^{239}Pu as a fissile nuclide (compared to ^{235}U). In fast reactors, the fission of fertile nuclides (heavy nuclei with an even number of neutrons) plays an important role. Depending on the fission cross section of such nuclei (^{238}U, ^{232}Th) on the kinetic energy of the neutron that caused the fission, there is a threshold at $E \approx 1.4$ MeV. In the spectrum of the thermal reactors, the contribution of the fission of fertile nuclides to the effective neutron multiplication coefficient is small: 5–7%. In fast reactors, 1/4–1/3 of the fertile nuclides are fissioned. The value of β for ^{238}U is 1.61%, for ^{232}Th is 2.28%. When used in a fast reactor mixed U-Pu-fuel containing dump uranium and plutonium with a high concentration of ^{239}Pu, the effective fraction of delayed neutrons will be significantly higher than when fission isotope-pure ^{239}Pu: about 0.36–0.38%, and when using ^{232}Th—even higher.

A certain role is played by fast neutron fission of some nuclei from among the minor actinides, for example, long-lived radioactive waste with an even number of neutrons: ^{237}Np and ^{241}Am, contained in small quantities (up to 5%) in the fuel of new-generation fast reactors to recycle this waste and protect the fuel cycle from proliferation. The value of β for ^{241}Am is 0.16%. The fission threshold of these nuclei is slightly more than 1 MeV. In the fast reactor spectrum, up to 1/3 americium nuclei can fission, which leads to a decrease in the average value of β and an increase in the potential danger of reactivity accidents. From the standpoint of the deterministic approach to exclude this assident, it is necessary to prevent the input of reactivity (ρ > β). To do this, it is necessary to limit (minimize) the reactivity reserves and, first of all, the fuel burnup reactivity reserve ($\Delta\rho_b$). For a homogeneous core, the conditions breeding ratio in core BRC = 1 and $\Delta\rho_b = 0$ are equivalent. In practical cases, the total reactivity margin can be limited to β.

2.3.2 The problem of positive void reactivity effect

The VRE is characterized by a strong spatial dependence. The most dangerous is the drainage of the central part of the core [4, 16]. In this case, the neutron leakage from the core is small, and, as a rule, the VRE is maximum. Draining the reflector leads to an increase in neutron leakage from the core, with the VRE < 0. During the transition to high-power reactors, the volume of the core increases, and the leakage of neutrons from it decreases, which leads to an increase in VRE. In high-power reactors with a traditional core layout, the local VRE, which is realized when its central part is drained, is positive and, as a rule, several times higher than the effective fraction of delayed neutrons.

There are several components of VRE associated with spectrum changes and neutron leakage, reduction of parasitic absorption, and changes in self-shielding factors [4]. The spectral component of the VRE and the leakage component are maximal in absolute value and opposite in sign. They are mainly and determine the VRE. From the point of view of VRE minimization, a coolant based on double-magic nuclei (e.g., lead-208), fuel and structural materials with a high neutron capture cross section are of interest.

2.4 Problems of oxide fuel use

1. When using dioxide fuel in fast reactors of medium and high power, there are some problems associated with the opposite role of the Doppler coefficient of reactivity in emergency situations of different types and, as a consequence, the contradictory requirements for the Doppler coefficient in the design of self-protected reactors [16–18].

2. When using dioxide fuel (including with a reduced proportion of oxygen), traditional fuel assemblies, and fuel pin grids, the maximum value of BRC < 1 and does not exceed 0.86. This is due to the elastic scattering of neutrons on oxygen nuclei and, as a consequence, a decrease in the energy of neutrons and a decrease in the average number of neutrons produced per fission. The low thermal conductivity and density of the dioxide fuel reduce the self-protection of the reactor. Low BRC requires significant (several times greater than β) reserves of reactivity burnup, which does not allow deterministically to exclude reactivity accidents (with the reactor is prompt supercritical). Monoxides do not have the necessary temperature resistance and are not used as fuel for nuclear reactors.

3. At operating temperatures, oxygen is released from plutonium oxide and migrates to the cladding of fuel pin [18]. In the presence of free oxygen, the chemical interaction of nuclear fuel, fission products (cesium, iodine, tellurium, bromine, selenium, antimony), and technological impurities (chlorine, fluorine, carbon dioxide) with the cladding of fuel pin reduces the strength and plasticity of the cladding, increases the rate of corrosion of the inner surface of the cladding. Developers of oxide fuel for LMFRs in the USSR, USA, and Great Britain faced the problem of oxygen yield at the dawn of the development of fast reactors. To "bind" free oxygen, a getter is needed. In Russia, the first experiments on the use of vibrocompacted MOX fuel with uranium getter (3–10% by weight) for LMFRs were started at JSC "VNIINM" (Moscow), then continued at JSC "SSC NIIAR" (Dimitrovgrad) [19, 20]. Research was carried out at the BOR-60 reactor since 1981, BN-350 since 1982, and BN-600 since 1987. The transition to such fuel in BN-600 did not solve the problem of corrosion of the inner surface of the cladding.

2.5 Modern directions of development of power LMFRs

As noted, the world is considering two main concepts of medium-power and high-power LMFRs: with sodium and lead coolant. As a possible heat carrier for use in low-power reactors, the eutectic alloy Pb-Bi [15, 21, 22] is considered. In the 1990s, South Korea, China, Japan, and other countries showed interest in fast lead-bismuth reactors. In Russia, projects of such low-capacity reactors are being developed [15, 22]. Bismuth is significantly more expensive than lead. The use of bismuth leads to the working time of the short-lived highly active isotope polonium-210 in the heat carrier. In the 1990s, systems for purifying heat carrier from polonium were actively developed during reactor operation [23].

Sodium coolant is technologically developed. In many countries, it is preferred in the development of LMFRs [15]. In Russia in 2016, the power unit of the BN-800 was put into power operation; the project of the BN-1200 power reactor is being developed [24, 25].

Lead has not yet been used as a coolant. According to the developers of BREST projects, their implementation is possible within the framework of existing Russian technologies [14]. The key works that initiated the development of the BREST direction can be considered articles [13, 26–28].

Works on the development of lead-cooled fast reactors are actively carried out in Russia (technical design and planned construction of BREST-OD-300, work on the concept design of MN fuel BREST-1200 is underway [27, 29]), European Union countries (concept projects ELFR, ELSY, LEADER, ALFRED with MOX fuel have been developed since 2006 [30]), and the USA by 2005–2007, and conceptual projects of STAR series of reactors of different target purpose with UN-fuel have been proposed [22, 30].

2.6 Problems of BREST concept during transition to high-power reactors

In terms of the possibility of avoiding severe accidents among fast reactors, the BREST-OD-300 project is most attractive [14]. It's a low-power reactor. It is well-known that it is much easier to ensure the safety of a low-power reactor. At reactor power increase (BREST-1200), there are problems with inherent safety provision. For example, the VRE is several times greater than β. The achievement of high economic efficiency of the NPP with BREST reactor may be hampered by the relatively high rate of corrosion-erosion of liquid lead on structural materials. If in BREST-OD-300 it is estimated at 39 kg of steel (for core only) per year [14, 31], for BREST-1200 it can be $39 \cdot 4 = 156$ kg/year. At the same time, economic estimates should correspond not to the cost of steel but to the cost of fuel assemblies, in which fuel elements will be depressurized.

The use of MN fuel will be accompanied by the release of nitrogen (from the PuN) and its migration to the cladding of fuel pin. In the presence of free nitrogen, the corrosion rate of the inner surface of the claddings increases. (The process is similar to the release of free oxygen from MOX fuel. The only difference is that the corrosion rate in the presence of nitrogen is less than in the presence of oxygen).

The absence of zirconium in reactor core allows deterministic elimination of steam-zirconium chemical reaction. However, hydrogen release is possible in other metal-based chemical reactions implemented at high temperatures. Chromium reacts with steam (water) in the absence of oxide film on the fuel element surface: $2Cr + 3H_2O \leftrightarrow Cr_2O_3 + 3H_2$. In this reaction, heat and free hydrogen are released. The surface of cladding must be carefully protected from possible contact with the steam involved in the core. (Chromium is present as part of the structural steel of fast reactors. The BREST uses a two-circuit cooling systems. Lead is primary circuit, steam is secondary circuit).

According to journalists, the interaction of nitrogen with lead (if there is a lead layer between the fuel and the cladding) can cause the formation of explosive lead azide. Lead does not react with nitrogen and such an event is highly unlikely. Lead azide production requires special conditions and reagents not present in BREST.

2.7 The considered emergency operation

As part of the deterministic approach to safety analysis [4], ATWS requires priority consideration. In fast reactors with liquid metal cooling, the whole set of emergency situations, including ATWS, is described (simulated) by perturbations of reactivity δρ, coolant flow rate δG, and coolant temperature at the core inlet δT_{in}. When designing LMFRs, the following emergency situations (commonly used abbreviations are specified) and their combinations correspond to various disturbances:

- $\delta\rho > 0$—TOP WS (transient overpower without scram).
- $\delta G < 0$—LOF WS (loss of flow without scram).
- $\delta G > 0$, $\delta T_{BX} < 0$—OVC WS (overcooling accident without scram).
- $\delta T_{BX} > 0$—LOHS WS (loss of heat sink without scram).
- Separately consider the loss of coolant accident without scram (LOCA WS).

Simulation of emergency modes LOF WS, TOP WS, LOHS WS, and OVC WS was performed using a code version of FRISS-2D and DRAGON-M [1]. DRAGON-M and MCU codes [32] are used to analyze LOCA WS.

2.8 Problem definition

Modern projects of fast reactors with sodium and lead coolant have some disadvantages and, therefore, not yet used reserves in terms of increased safety. This is particularly evident in the transition to high-capacity power reactors (e.g., BN-1200, BREST-1200). There are opportunities to improve the fuel, coolant, and structural materials used. The greatest effect in improving the safety of the next-generation reactors should be expected with the complex (simultaneous) use of improved core materials.

This chapter offers some ideas for improving fuel, coolant, and structural materials to improve reactor self-protection.

3. Lead-208-based coolant

Among the realistic liquid metal coolants of fast reactors are sodium, lead, and eutectic alloy of lead and bismuth. **Table 1** shows data on the prevalence of lead, bismuth, and sodium, their world production, and world reserves (according to data [33]).

Natural lead is intended to be used in lead-cooled reactors. The isotopic composition of such lead (^{nat}Pb: 1.4% ^{204}Pb – 23.6% ^{206}Pb – 22.6% ^{207}Pb – 52.4% ^{208}Pb) is obtained by averaging all known deposits (about 1500). The isotopic composition of lead in different fields can vary significantly (see **Table 2**).

The large differences in the isotopic composition of lead from different deposits open up great opportunities for optimizing the composition of lead coolant without the use of expensive isotope separation technology. By mixing lead from different deposits, it's possible to obtain a coolant with a given concentration of stable

Element	Prevalence		World production, tons/year	World reserves, tons
	10^4%	Place		
Na	23,000	6	2×10^5 (metal), 2.9×10^7 (carbonate), 1.68×10^8 (salts)	Are almost not limited
Pb	14	37	4.1×10^6	8.5×10^7
Bi	0.048	70	3×10^3 (associated with lead and copper)	There are no data

Table 1.
World production, reserves, and prevalence of certain elements in the Earth's crust.

Ore	^{204}Pb	^{206}Pb	^{207}Pb	^{208}Pb
Monocyte of pegmatites of granite thorium ores	0.01–0.076	0.89–26.43	0.35–4.11	69.15–97.74
Uranitol of granite of uranium ore gneiss	0.12–0.15	85.53–85.81	10.49–10.54	3.58–3.69
Zircon of a pegmatite pink, granites, monocyte, magmatite	0.101–0.320	46.71–75.58	7.74–12.93	12.48–45.40
Lead and polymetallic ore galenite	1.09–1.61	18.64–25.17	21.36–30.80	49.30–52.49
Deep sea and Pacific Ocean	1.34	25.43	21.11	52.12
Natural lead (^{nat}Pb)	1.4	23.6	22.6	52.4

Table 2.
Isotopic composition of lead of different deposits, % (wt) [16].

isotopes, i.e., with given nuclear-physical properties. If it's necessary to minimize VRE, thorium lead with a high concentration of ^{208}Pb (decay product of ^{232}Th) is preferable, and if it is necessary to minimize the production of ^{210}Po, uranium lead with a high concentration of ^{206}Pb (decay product of ^{238}U), and small impurities of ^{207}Pb (decay product of ^{235}U), as well as non-radiogenic lead ^{204}Pb are required. The use of uranium and non-radiogenic lead leads to an increase in VRE.

Figures 1 and **2** present statistics on the isotopic composition of lead in the monocytes of the Ukrainian shield. The age of the breed is about 2 billion years. An analysis of 49 samples was carried out: along the abscissa axis is the sample number according to [34].

Figure 3 presents the results of the MCU calculation of the void and density effects of reactivity (Δρ) in a high-power BREST reactor. In **Figure 3(a)**, the dependence of VRE on the content of ^{208}Pb in the coolant is given. **Figure 3(b)** shows the dependence of the effects of reactivity on the density of a coolant based on lead of polymetallic ores. Various drainage scenarios are considered: (1) the entire reactor (when calculating the density effect—the change in the density of the coolant in the entire reactor); (2) the core and the lower reflector (or a change in the density of lead); and (3) drainage of the core. The implementation of VRE involves the complete loss of coolant (drainage of the reactor, core or part thereof), the implementation of the density effect involves a change in the density of the coolant.

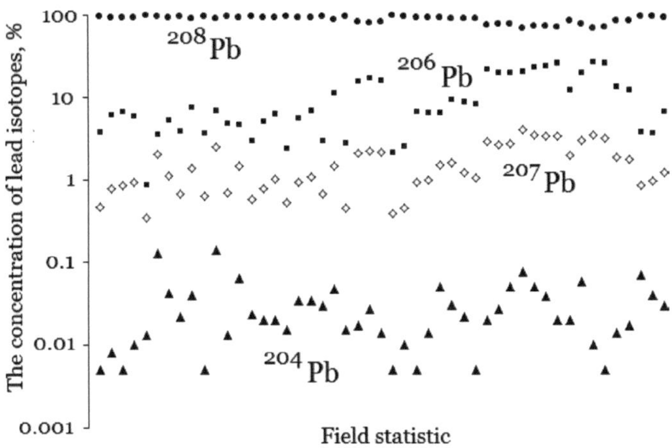

Figure 1.
The concentration of lead isotopes in different samples of monazite.

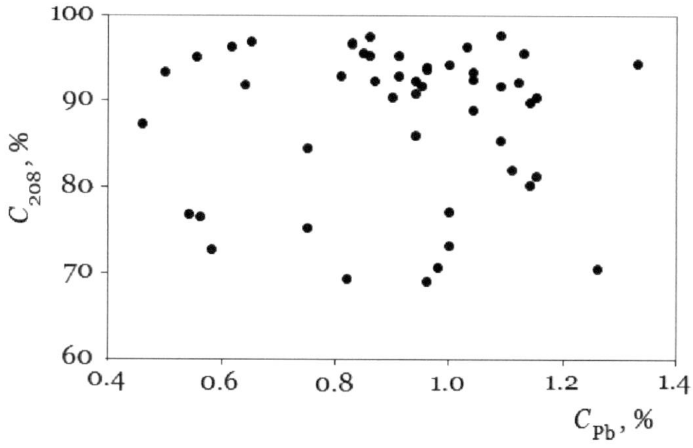

Figure 2.
The dependence of the content of the isotope ^{208}Pb in lead (C_{208}) on the lead content in monazite (C_{Pb}).

Table 3 shows the VRE values (calculations according to the MCU code) for various scenarios of drainage of the BREST reactor. For comparison, the last line shows the VER values for the sodium-cooled reactor. The numbers indicate drainage scenarios (see **Figure 3**).

Mineral, ore, isotope	The isotopic composition of coolant, % (wt)				VRE (scenarios), $		
	^{204}Pb	^{206}Pb	^{207}Pb	^{208}Pb	1	2	3
Monazite	0.013	0.89	0.35	97.74	−31.7520	−15.1263	11.6096
Zircon	0.226	75.58	11.54	12.48	−14.2736	0.4048	22.6550
Uraninite	0.150	85.53	10.64	3.69	−12.8460	1.6599	25.0906
^{nat}Pb	1.4	23.6	22.6	52.4	−20.5426	−4.2157	−4.2151
^{204}Pb	0	0	0	100	−32.4267	−15.16023	11.2648
^{nat}Na	^{23}Na (100%)				−4.3644	2.5481	19.3386

Table 3.
VRE values for various drainage scenarios (LOCA WS).

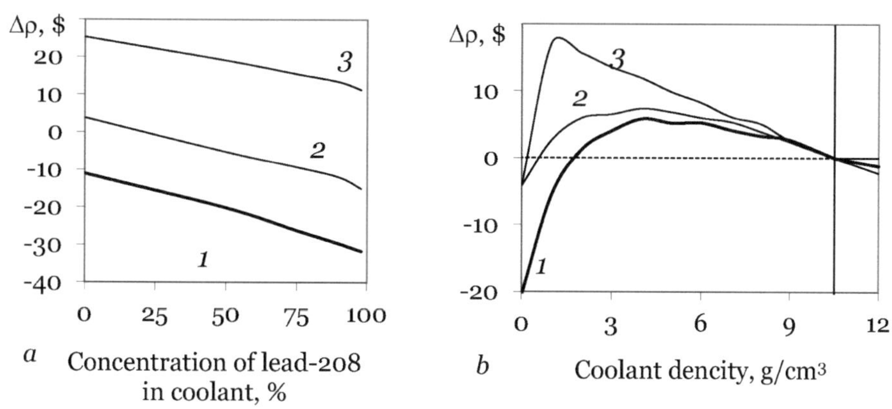

Figure 3.
Studies of the void and density effects of reactivity in a high power BREST reactor, (a) the dependence of VRE on the content of ^{208}Pb in the coolant is given, (b) the dependence of the effects of reactivity on the density of a coolant based on lead of polymetallic ores.

When using lead with a high concentration of ^{208}Pb as a coolant, the potential risk of reactivity accidents is reduced. The neutron absorption cross section of ^{208}Pb in the spectrum of the BREST reactor is 10^{-3} b (1 b = 10^{-28} m^3), that is, approximately three times lower than ^{nat}Pb. The core of such a reactor is characterized by a large volume fraction of lead and a small fraction of fuel, which increases the role of the coolant in changing the neutron macroscopic absorption cross section Σ_a by core materials. When switching to thorium lead, the Σ_a decreases, which leads to an increase in the lifetime of prompt neutrons and limits the prompt supercritical when reactivity is introduced that is greater than β. When switching to uranium lead (^{206}Pb), the Σ_a increases (the neutron absorption cross section of ^{206}Pb is more than three times higher than ^{nat}Pb); as a result, the potential danger of reactivity accidents (the reactor is prompt supercritical) increases.

One of the ways to increase the corrosion resistance of structural materials is the use of technological additives (inhibitors and deoxidants) to the coolant. The possible inhibitors are characterized by a high absorption cross section and quickly burn out in a neutron field, forming slags in the coolant. All metallic impurities contained in liquid lead (except bismuth), and all metallic components that may be present in lead as a result of corrosion and erosion of structural materials, have a lower electrode potential than lead, i.e., these impurities play the role of deoxidizing agents. The deoxidizing agent may also be an alkali metal. Small additives of lithium or potassium form a eutectic with lead, reducing the freezing point of the coolant. Small sodium additives play the role of an alloying additive to lead, increasing the freezing point of the coolant. Lithium is the strongest reducing agent—the best getter of free oxygen. Experiments conducted at the JSC "SSC RF-IPPE" (Obninsk, Russia) show that the cardinal way of changing the initial oxidizing properties of heavy coolants is the use of additives of metal deoxidizing agents that can reduce the level of oxidative potential of the melt. Thus, a technological additive of 1.8% (wt.) potassium to lead allows the formation of a eutectic alloy with oxygen activity five orders of magnitude lower than that of a pure lead melt at temperatures of 500–550°C.

With a decrease in the mass number of nuclei, the role of elastic moderation increases, and the void and density effects of reactivity increase. The relatively high VRE with the use of a coolant based on the eutectic Pb-^{7}Li alloy can be reduced by switching from natural lead to ^{208}Pb in the composition of this alloy (see **Figure 4**). However, in this case, VRE is significantly higher than when using lead polymetallic ores.

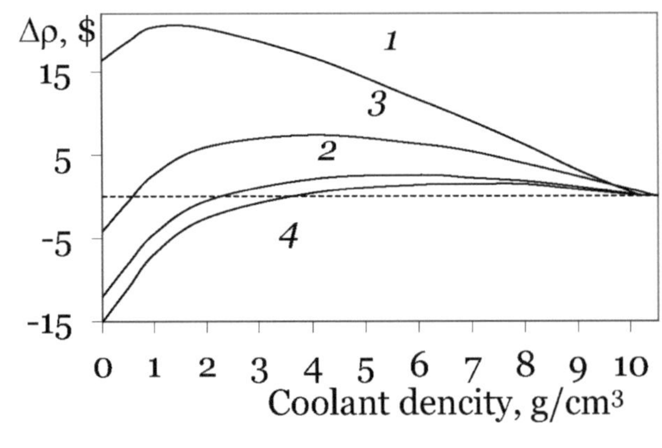

Figure 4.
Change in the reactivity effect Δρ depending on the density of the coolant in the core and the lower reflector during drainage of the BREST high-power reactor from the bottom up (involvement of bubbles in the core), (1) eutectic ^{7}Li-^{208}Pb; (2) natPb; (3) eutectic natK-^{208}Pb; (4) ^{208}Pb.

More detailed results of studies on the use of a coolant based on lead extracted from thorium ores are presented in [16, 35–38]. To ensure acceptable VRE in high-power reactors, it is not necessary to use isotopically pure lead-208. It is enough that the content of ^{208}Pb in the lead coolant is at least 75–80%. This makes it possible to use almost any thorium deposit for lead mining.

4. Construction materials

The greatest successes have been achieved in the field of creating new structural materials. For fast sodium reactors, specialists from Russia, the USA, Japan, China, France, and Ukraine consider radiation-resistant heat-resistant materials based on dispersion—hardened by thermally stable nanooxides Y_2O_3, Y_2O_3-TiO_2 or Al_2O_3 (3–5 nm) ferritic-martensitic steels with good strength and mechanical properties capable of working at neutron fluences (with kinetic energies of more than 0.1 MeV) up to 2×10^{16} cm^{-2} s^{-1} to damaging doses of 160–180 sleep at temperatures of 370–710°C (see [39–41] and links to these works).

While ensuring a good neutron balance in the core due to the use of innovative fuels, coolant, and structural materials, two advantages of additional absorbers in structural materials can be distinguished.

Firstly, the fast neutron absorption cross section of structural materials that cannot leave the core when it is drained favorably affects the development of the LOCA emergency: they contribute to the reduction of VRE. In the manufacture of structural materials, the content of alloying additives is regulated by regulatory documents. The content of alloying additives can vary in a given range. A negative factor may consist in a change in the sign of VRE depending on the content of alloying additives in the composition of structural steel (see **Figure 5**).

Secondly, the removal from the core of erosion and corrosion products (of structural materials) that strongly absorb neutrons helps to prolong the campaign of the reactor, increasing reactivity, similar to the action of burnable absorbers in thermal reactors (see **Figure 6**). The negative factor is a decrease in the thickness of the cladding of the fuel rods and slagging of the contour (increase in erosion rate) by corrosion and erosion products that have not settled on the "cold" section of the coolant circulation path. As a result, the processes of corrosion and erosion of structural steels contribute to solving the problem of minimizing the reactivity

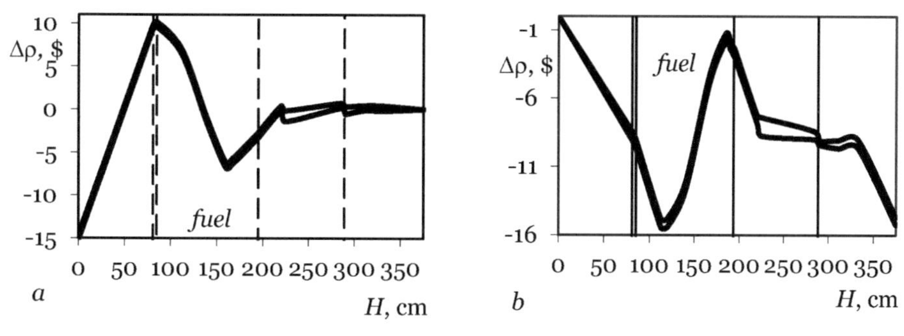

Figure 5.
VRE in a BREST reactor of high power under conditions of uncertainty in the concentration of alloying additives in structural steel within the permissible (MCU). H—The boundary of the medium "lead-vacuum." conservative assessment: Horizontal neutron leakage is equal to zero; 100% replacement of lead by vacuum are not taken into account, (a) the drainage boundary moves from top to bottom (lead discharge, in BREST the event is ruled out deterministically), (b) the drainage boundary moves from bottom to top (involvement of bubbles in the core during depressurization of the lead-water heat exchanger tubes).

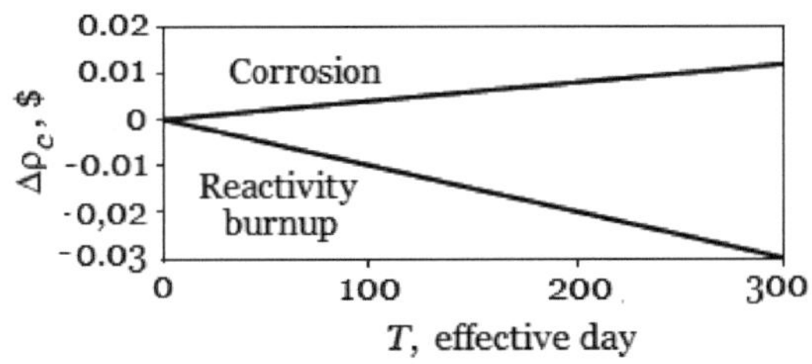

Figure 6.
Change in the reactivity Δρc for microcampaign T as a result of corrosion of the structural steels and the fuel burnup.

margin for fuel burnup. In BREST projects it is proposed to use steel without nickel but with increased silicon content [14]. Silicon promotes the formation of protective oxide films on the outer surfaces of cladding of fuel element.

The calculated analysis made it possible to conclude on the possibility of using tungsten coatings of fuel element cladding from the outside and inside. This will lead to the improvement of reliability and safety of BREST reactors without deterioration (and possibly with improvement) of economic characteristics of NPPs with such reactors [16, 42, 43]. The application of tungsten coverings will allow to reduce the corrosion rate and erosion in liquid lead and will lead to the deterioration in neutron balance (and perhaps, reactor costs); minimization will be promoted by the VRE (large absorption cross section of fast neutrons) and will open possibilities of use of the lead more polluted by impurity.

Tungsten does not react with nitride fuel at temperatures below 1485°C. This temperature is not achieved even in ATWS [14, 16]. Tungsten interacts weakly with lead. Conditions for the formation of lead ($PbWO_4$) tungsten in the reactor core are poor: insufficient oxygen, high-energy neutrons break chemical bonds.

Tungsten in combination with chromium improves mechanical properties of steel [44]. Due to the small (compared to iron) values of the cross section of neutron interaction with tungsten nuclei, accompanied by the yield of gaseous products (less than 10^{-2} b), and the high threshold of such reactions (10 MeV), tungsten does not swell and does not become brittle at high fluxes of fast neutrons.

Tungsten coatings produced by plasma spraying have a layered structure, do not crack, or peel even when bent [45]. Tungsten plasma spraying technology is well developed. The dense (nonporous) tungsten film completely covers the steel surface. Corrosion and erosion resistant coatings in a wide range of temperatures (higher than ATWS temperatures) in liquid metal melts, including under conditions of additional abrasive wear (practically excluded by tungsten coatings).

Like silicon, tungsten contributes to the formation of a protective oxide film on the surface of the shell (to a lesser extent because it holds oxygen slightly worse: bond energy SiO = Si + O is 8.24 ± 0.10 eV, WO = W + O is 6.94 ± 0.43 eV [46]). Due to the leakage of the circuit, oxygen will always be present in the lead. The oxide film (WO_2, W_4O_{11}, WO_3) holds well (up to 923°C) and serves as an additional naturally formed protective coating.

The tungsten content of the sprayed layer is about 1%. As a result, the increase in the cost of the structural material may not be essential. Among the neutron-physical aspects of the use of tungsten-sprayed cladding, the following are noted. The deterioration of the neutron balance may not be essential when thin coatings are used. (At that, it is possible to reduce thickness of cladding).

The MCU code calculations showed that an increase in the proportion of tungsten in the structural material from 0 to 100% leads to a marked decrease in VRE: by 1.76 $ with hypothetical drainage of the core and the lower end reflector and

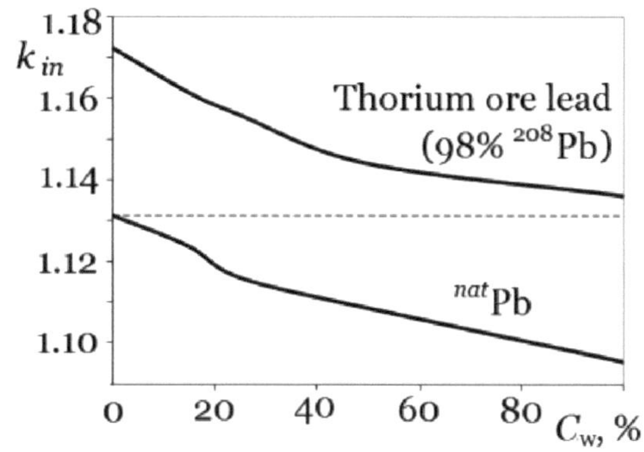

Figure 7.
The dependence of k_{in} on C_W.

when 98% ^{208}Pb in the lead coolant. The most realistic and most dangerous scenario of such an emergency situation is not related to the complete drainage of the reactor but to the involvement of bubbles in the core during depressurization of the pipes of the "lead-steam (water)" heat exchanger. At that, reactivity effect is maximum at the reduction of lead density in the core and lower end reflector from 10.5 to 7.0 g/cm^3 [35]. In such a scenario, the density effect of reactivity is $0.06 when used as a structural material of corrosion-resistant steel and minus $ 0.86 when switching to tungsten cladding. To obtain a conservative estimate, an infinite array of fuel elements is chosen, formally corresponding to the reactor of infinite power.

Figure 7 shows the dependence of the infinity multiplication factor kin on the mass content of tungsten C_W in the structural steel of the BREST-OD-300 core. It is possible to fully compensate for neutron losses due to their absorption by tungsten nuclei by increasing the proportion of ^{208}Pb in the heat carrier even when using shells made entirely of tungsten.

The other way is connected with self-organizing coatings of internal surface of fuel element cladding. It is proposed to place liquid lead saturated with zirconium or liquid eutectic alloy 97.53% (wt.) Pb, 2.25% Mg, and 0.2% Zr [47]. When the reactor is operating, a protective coating based on zirconium nitride is spontaneously formed on the inner surface of the cladding. Self-healing of accidental damages of coating is provided.

5. Accident-tolerant fuel

5.1 General provisions

Fast reactors of the new generation are focused on the use of ceramic mixed uranium-plutonium fuel. The immediate prospects are associated with the use of mixed oxide (MOX). A more promising in terms of possible exceptions of severe accidents in NPP can be considered mixed mononitride uranium-plutonium fuel (MN fuel, mixed nitride). It is proposed to use in BREST reactors with lead cooled and, in the future, perhaps in power sodium-cooled reactors.

Nanotechnology, in relation to the creation of new materials for nuclear technology, is widely used since the early 1960s.

Thanks to the introduction of nanotechnology the unique fuel materials, characterized by high density and thermal conductivity, can be obtained. The task of creating high-performance nuclear fuel (mostly fast reactors), which is achievable for the high burnup, isolated two problems [39]:

- Creating a coarse structure with a given porosity (MOX with nanoadditives for BN-type reactors [40]), capable of holding gaseous and volatile corrosive fission products, preventing their migration along the grain boundaries to the cladding in order to slow down corrosion of the inner surface of the cladding.

- The activation process of sintering fuel pellets due to the nanoadditives [40].

The porosity of the ceramic nuclear fuel consisting of micrograins is approximately 25%. When making fuel from nanopowder, it is possible to significantly reduce porosity (up to 5–10%). For example, in order to increase the density of MN fuel to 95% of the theoretical density by reducing porosity, the inventors [48, 49] propose a fuel based on MN nanopowder. In terms of increasing reactor power, reducing porosity is equivalent to increasing the volume of fuel in the core or increasing the volume of the core.

It is known that, compared to ceramic fuel (MOX, MN, etc.), cermet fuel has some advantages. It is more attractive due to increased thermal conductivity and density, increased BRC (up to 1). The main disadvantage preventing the use of such a fuel is the reduction of the melting point. The use of a nanopowder of metallic uranium in conjunction with ceramic fuel micrograins will provide the same advantages, eliminating the disadvantage. Ideally, the pores between the micrograins can be filled with nanopowder. In this case, the melting point of the fuel will be determined by the melting point of the ceramic (MOX, MN, etc.). The nanopowder in the fuel composition can melt in emergency modes; there will be droplets of molten metal between the ceramic microloans, which will have practically no impact on the reactor safety. As a result, it is possible to obtain fuel having high thermal conductivity and density, which contributes to self-protection of the reactor. When using such fuel, the condition BRC = 1 is easily achievable.

At the beginning of the twenty-first century, in Russia fast reactors with sodium cooling was considered vibrocompacted MOX-fuel with uranium (up to 10% by weight) getter of free oxygen [19, 20]. Since it is difficult to control the uniformity of mixing MOX and uranium metal powders in a relatively large volume of fuel element (in case of vibration compaction), it makes sense to switch to small volumes (pellet fuel) [50, 51].

The finer the uranium powder particles, the better it exhibits getter properties. Nanodisperse powder is one of the best free oxygen getter. The ideal getter is a powder ground to size when any atom in the nanoparticles can be considered surface. (The best getter is thorium [46]. However, its use will require a reorientation of Russian nuclear fuel cycle enterprises, which is not economically feasible).

5.2 ATF based on ceramics and beryllium (a new look at the old concept)

The free oxygen getter may be a material having a high chemical affinity for oxygen (see **Table 4**). The first steps towards solving the problem of free oxygen binding were taken in the late 1950s and early 1960s [52]. Beryllium (and Be_2O) can also serve as a getter. The possibility of using oxide fuel with beryllium additives was considered at the dawn of the development of fast reactor technologies [4]. This was abandoned due to the decrease in BR.

Homogeneous placement of metal beryllium in MOX or MN fuel helps to solve the problem of corrosion of the inner surface of cladding and increase self-protection [51]. The spectral component of the VRE for an endless reactor with MN fuel and MN-5%Be fuel is $23.6 and 12.7, respectively [51]. The Doppler reactivity coefficient for infinite array of fuel elements is -7.99×10^{-6} and -2.06×10^{-5} K^{-1}, respectively.

Getter (bold type)	*E*, eV	Getter (bold type)	*E*, eV
ThO = **Th** + O	8.59 ± 0.22	TiO = **Ti** + O	6.85 ± 0.09
TaO = **Ta** + O	8.37 ± 0.43	ZrO_2 = **ZrO**+ O	6.68 ± 0.48
ThO_2 = **ThO** + O	7.89 ± 0.30	VO_2 = **VO** + O	6.51 ± 0.30
UO = **U** + O	7.81 ± 0.17	UO_3 = **UO2** + O	6.46 ± 0.30
ZrO = **Zr** + O	7.81 ± 0.43	VO = **V** + O	6.29 ± 0.43
YO = **Y** + O	7.37 ± 0.13	Be_2O_2 = **Be2O** + O	5.94 ± 0.65
UO_2 = **UO** + O	7.37 ± 0.30	BeO = Be + O	4.60 ± 0.13

Table 4.
Chemical bond energy E for some getter and oxides formed (as reported by [46]).

The increase in the thermal conductivity of the fuel (due to beryllium) leads to a decrease in the temperature of the fuel. As a result, the nature of the change in maximum fuel temperatures in medium or high power reactors (BN-800, 1200, 1800) in LOF WS and TOP WS is same. To increase self-protection against these accidents, a large absolute negative Doppler reactivity coefficient is required.

Additives of beryllium to MOX fuel contribute to the minimization of corrosion rate of fuel element shells from inside, the reduction of VRE, the elimination of known contradiction in requirement for Doppler reactivity coefficient in emergency modes LOF WS and TOP WS in high-power reactors, and the improvement of reactor self-protection. At the same time, beryllium supplements reduce BRC and BR. It's a major flaw.

5.3 ATF based on MOX and uranium nanopowder

For MOX-fueled reactors, it is possible to offer pellet fuel based on fine-grained MOX and nanopowder U. Fuel density (at 18% U) under normal conditions of about 12 g/cm^3, thermal conductivity—12 $W\ m^{-1}\ K^{-1}$. Increasing the density and thermal conductivity of MOX-U fuel with increasing U fraction leads to changing the role of Doppler reactivity coefficient in providing self-protection against accidents of LOF WS type. As a result, the role of Doppler reactivity coefficient in providing self-protection against accidents of type TOP WS and LOF WS becomes the same (or almost the same), which helps to resolve the conflict situation when optimizing the core layout. This makes it possible to significantly increase reactor safety.

Figure 8 shows the results of modeling LOF WS mode (FRISS-2D code) in LMFR using different fuels. The relationship between maximum temperature of fuel and heat carrier is presented when using pellet MOX fuel in BN-800 (a), MN fuel in reactor of type BREST-OD-300 [14] (b), and BN-800 (c). The thin line corresponds to the conservative approximation: constant coolant temperature at the core inlet in emergency mode. **Figure 8(d)** shows the time dependence of the maximum temperature of the MOX-based pellet fuel and uranium nanopowder in the TOP WS and LOF WS modes and when they are superimposed (dotted line in **Figure 8(d)**). For correct comparison of variants at use of different fuel, the same fuel assemblies and grids of the core BN-800 [53] are accepted.

LOF WS is initiated by de-energizing of all main circulation pumps of the primary circuit at coastdown time 30 s (for BN-800) or by reduction of flow rate in 20 s (for BREST-300). TOP WS is initiated by reactivity input $0.3 for 10 s.

In order to increase the self-protection of reactors against the TOP WS accident, it is necessary to increase the negative Doppler reactivity coefficient by absolute value. In LOF WS mode, when MOX fuel is used, the maximum fuel

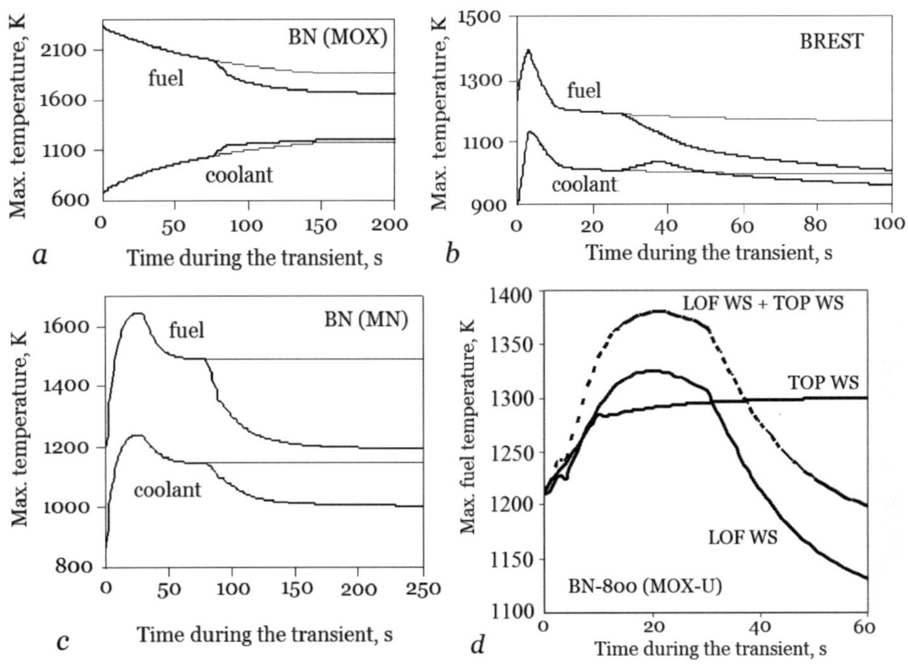

Figure 8.
Time dependence of pellet fuel temperature in LOF WS mode for BN-800 reactors with MOX fuel (a), БРЕСТ-300 (b), and БН-800 (c) with MN fuel, BN-800 with MOX-U fuel (d).

temperature decreases with time. At the same time, it is possible to start boiling sodium. In order to increase self-protection, it is necessary to modulo reduce the negative Doppler reactivity coefficient. The use of vibration-compressed MOX fuel with uranium getter [19, 20] does not significantly change these dependencies, and the role of Doppler reactivity coefficient in TOP WS and LOF WS modes remains the opposite.

When using MN fuel in LOF WS mode, an intermediate scenario between metal and oxide fuel is implemented. To minimize maximum temperatures, the negative Doppler reactivity coefficient must be increased modularly, as well as to ensure self-protection against TOP WS accidents. If the maximum fuel temperature is reduced to less than the nominal value, it is necessary to modulo reduce the negative Doppler coefficient, but due to significant temperature reserves in this case, the role of Doppler coefficient has no determining value. This is the pattern (**Figure 1(d)**) observed in the emergency mode LOF WS when using a tablet fuel based on fine-grained MOX and nanopowder U (18% wt.). At its low U content, the scenario of emergency mode development is the same as when using tablet or vibrocompressed MOX fuel with finely dispersed (100 μm) U-getter. At high U content (up to 50% by weight), the nature of the change in maximum fuel temperature over time in the LOF WS mode is similar to that shown in **Figure 1(b)**. At mass content of uranium nanopowder, 18% fuel is slightly lower than the tablet MN in density and 1.5 times in thermal conductivity. As a result, when switching to MOX-U, the maximum fuel temperature is significantly lower than when using a tablet MOX fuel and approximately the same as the MN fuel (with larger fuel pellet diameter). The nature of the change of maximum temperatures in emergency modes TOP WS and LOF WS in fast reactors of medium and high power with MN fuel and fuel based on fine-grained MOX and nanopowder U are the same. In both cases BRC = 1.

The LOHS WS and OVC WS are not dangerous.

The temperature of MOX-U fuel even in ATWS is lower than the oxide when the reactor operates at rated power. This helps to reduce oxygen release from the Pu_4O_7 in nominal and emergency modes of reactor operation.

When the temperature increases (in emergency modes), the MOX-U fuel can contain nanodroplets of liquid metal between MOX grains. By neutron balance this fuel is closer to the pellet monoxide or MN, by the melting point—closer to UO_2-Pu_4O_7. So, when using nanotechnologies in the manufacture, MOX-U-fuel opens up large "reserves," approaching it in properties to MN.

5.4 ATF based on MN and uranium nanopowder

Further the improvement of fuel characteristics while maintaining high melting temperature is possible at transition to fuel based on MN nanopowder (100–500 nm). To achieve a density of 85–95% theoretical, the required sintering temperature 1800–1900°C and a 10 time of 10–30 hours. At the same time, up to 15% plutonium is evaporated from the fuel [48, 49]. When using such fuel, the nature of time dependence of the maximum temperatures in emergency modes does not change qualitatively.

The best and cheaper alternative is fuel based on micrograins MN and nanopowder U. The beginning of MN decomposition by 450–500°C exceeds the sintering temperature of fuel tablets during their manufacture. Significant improvement in reactor performance and safety can be expected with MN-U fuel due to high density and thermal conductivity, equal role of Doppler reactivity coefficient in TOP WS and LOF WS, and neutron balance improvement (BRC = 1). By neutron balance and nature of emergency modes, MN-U fuel is close to U-Zr and U-Pu-Zr [11] but much higher temperature.

The most relevant is the possibility of using fuel based on fine-grained MN and nanopowder U in lead-cooled reactors with a lead layer between the fuel and the cladding. The chemical bond energy in UN and PuN is markedly lower than in oxides (for UN, 5.464 ± 0.217 eV; PuN – 4.857 ± 0.651 eV; UO – 7.805 ± 0.174 eV; UO2 = UO O – 7372 ± 0.304 eV; PuO – 7.459 ± 0.217 eV [46]). Although the operating temperature of MN fuel is much lower than MOX, nitrogen can exit MN. The absorption of nitrogen from MN fuel by metal uranium can slow corrosion of internal surface of cladding.

6. Conclusions

There is potential to further improve LMFR reliability and safety within the existing technologies.

In order to ensure acceptable VRE in lead-cooled LMFR of high power, the minimum content of isotope ^{208}Pb in the lead heat carrier should be about 75–80%. This is typical of thorium ores.

The use of tungsten coatings of fuel element cladding helps to reduce VRE, eliminate reaction of chromium with steam in emergency situations, and exclude hypothetical possibility of lead azide formation. With the simultaneous use of tungsten coatings and heat carrier based on lead extracted from thorium ores, it is possible to ensure a good balance of neutrons in the core.

The potential and reserves of MOX fuel in ensuring the safety of large LMFR with sodium or lead coolant have not yet been exhausted. Pellet fuel based on fine-grained MOX and nanoscale U has no advantage over pellet MN fuel.

The prospects for LMFR development may be related to the use of fuel based on fine-grained MN and nano-sized metal uranium powder.

Uranium nanopowder additives will not only increase the density and thermal conductivity of nuclear fuel, which is important in the problem of reactor safety, but also significantly improve the neutron balance (by directing "excess" neutrons to ensure safety: achievement of condition BRC = 1 even on oxide fuel), almost completely eliminate the conflict character in the requirement for Doppler reactivity coefficient in terms of self-protection against ATWS in large LMFR, which opens up wide possibilities for the optimization of core layout without deterioration of safety.

The complex use of the proposed innovations allows to ensure the safety of large LMFR. So, pellet MN fuel with nanopowder of metallic uranium, a coolant based on lead extracted from thorium ores, and steel claddings of fuel elements with a tungsten coating together can significantly increase the safety of the reactor. This is accident-tolerant materials for LMFR.

The results presented can be easily summarized into other types of reactors. The research can be useful in the development of new-generation reactors and the improvement of technical solutions to existing nuclear reactors.

Conflict of interest

The author declares that he has no conflict of interest.

Thanks

The author is sincerely grateful to the Head of the Department of Physics at the Bauman Moscow State Technical University, Doctor of Physical and Mathematical Sciences, Professor, corresponding member of the Russian Academy of Sciences, A.N. Morozov, for his comprehensive assistance and support of research.

Author details

Viacheslav Sergeevich Okunev
Department of Physics, Bauman Moscow State Technical University, Moscow, Russia

*Address all correspondence to: okunevvs@bmstu.ru

References

[1] Kuzmin AM, Okunev VS. Using Variation Methods for Solving Problems of Ensuring and Justifying the Natural Safety of Fast Neutron Reactors. Moscow: MEPhI; 1999. p. 250

[2] Okunev VS. Comparative analysis of safety of fast reactors cooled with alloys of liquid metals. News of higher educational institutions. Nuclear Power. 2001;**1**:57-64

[3] Okunev VS. Comprehensive comparative analysis of the use of liquid metals and their alloys for cooling fast reactors. In: Proceedings of Innovate Design and Technologies of Nuclear Power. 3-Ht International Scientific and Technical Conference. Vol. 1. Moscow: JSC "NIKIET" Publ; 2014. pp. 291-302

[4] Waltar AE, Reynolds AB. Fast Breeder Reactors. New York, Oxford, Toronto, Sydney, Paris, Frankfurt: PERGAMON PRESS; 1981. p. 853

[5] Golden GH et al. Evolution of thermal-hydraulics testing in EBR-II. Nuclear Engineering and Design. 1987;**101**:3-12. DOI: 10.1016/0029-5493(87) 90145-2

[6] Messich NC et al. Modification of EBR-II plant to conduct loss-of-flow-without-scram tests. Nuclear Engineering and Design. 1987;**101**: 13-24. DOI: 10.1016/ 0029-5493(87) 90146-4

[7] Lahm CE et al. Driven fuel qualification for loss-of-flow and loss-of-heat sink tests without scram. Nuclear Engineering and Design. 1987;**101**:25-34. DOI: 10.1016/0029-5493(87) 90147-6

[8] Lehto WK et al. Safety analysis for the loss-of-flow and loss-of heat sink without scram tests in EBR-II. Nuclear Engineering and Design. 1987;**101**:35-44. DOI: 10.1016/0029-5493(87) 90148-8

[9] Mohr D et al. Loss-of-primary-flow-without-scram tests: Pretest predictions and preliminary results. Nuclear Engineering and Design. 1987;**101**:45-56. DOI: 10.1016/ 0029-5493(87) 90149-X

[10] Feldman FE et al. EBR-II unprotected loss-of heat sink predictions and preliminary test results. Nuclear Engineering and Design. 1987;**101**:57-66. DOI: 10.1016/0029-5493(87) 90150-6

[11] Meneghetti D, Kucera DA. Reactivity feedback components of a homogeneous U10Zr-fueled 900 MW (thermal) liquid-metal reactor. Nuclear Technology. 1990;**91**:139-145

[12] Van Tuyle GJ et al. Examine the inherent safety of PRISM, SAFR, and the MHTGR. Nuclear Technology. 1990;**91**:185-202

[13] Adamov EO et al. Conceptual design of BREST-300 Lead-cooled fast reactor. In: Proceedings of Int. Conf. On Advanced Reactor Safety (ARS-94); 17-21 April 1994. Vol. 1. Pittsburg, USA; 1994. pp. 509-515

[14] Adamov EO, Orlov VV, editors. BREST-OD-300. Moscow: NIKIET; 2001. p. 315

[15] IAEA-TECDOC-1289. Comparative Assessment of Thermophysical Characteristics of Lead, Lead-Bismuth and Sodium Coolants for Fast Reactors. Vienna, Austria; June 2002. p. 65

[16] Okunev VS. The Basis of Applied Nuclear Physics and Introduction into Physics of Nuclear Reactors. Series "Physics in Technical University". Moscow: Bauman Moscow State Technical University; 2015. p. 536

[17] Khammel H, Okrent D. Reactivity Coefficients in Large Fast Power Reactors. Argonne National Laboratory: American Nuclear Society Publ; 1970. p. 310

[18] Okunev VS. About a role of Doppler coefficient of reactivity in safety of reactors on fast neutrons. In: Proceedings of the 12th Seminar on Problems of Reactors Physics "Physical Problems of Effective and Safe Use of Nuclear Materials"; 2-6 September 2002. Moscow: MEPhI; 2002. pp. 171-173

[19] Grachev AF et al. Experience and prospects of using fuel elements based on vibrocompacted oxide fuel. In: Proceedings of the International Conference "Nuclear Power and Fuel Cycles". Moscow - Dimitrovgrad; 2003. (CD-ROM. Section 1. Report No. 31)

[20] Maershin AA et al. Atomic Energy. 2001;**91**(5):378-385

[21] IAEA TECDOC-1569. Liquid Metal Cooled Reactors: Experience in Design and Operation. Vienna, Austria: IAEA; 2007. p. 263

[22] IAEA-TECDOC-1536. Status of Small Reactor Designs without on-Site Refuelling. Vienna, Austria: IAEA; 2007. p. 859

[23] Gromov BF et al. Use of melts of lead-bismuth eutectic and lead as nuclear-power-plant coolants. Atomic Energy. 1992;**73**(1):19-24

[24] Belov SB et al. Ensuring the operation of the BN-1200 core in equilibrium when using nitride fuel and MOX fuel. In. Proceedings of Innovate Design and Technologies of Nuclear Power. 3-ht International Scientific and Technical Conference; 7-10 October 2014. Moscow: JSC "NIKIET" Publ.; 2014. 1. pp. 47-53

[25] Vasilyev BA et al. Innovative Design of the BN-1200 Area. In: Proceedings of Innovate Design and Technologies of Nuclear Power. VI International Scientific and Technical Conference. 27-30 September 2016. Moscow: JSC "NIKIET" Publ. 1. pp. 31-41

[26] Orlov VV et al. Non-traditional concepts of nuclear power plants with natural safety (new nuclear Technology for the Next Stage Large-Scale Production of nuclear power). Atomic Energy. 1992;**72**(4):317-329

[27] IAEA-TECDOC-1348. Power Reactor and Sub-critical Blanket Systems with Lead and Lead-Bismuth as Coolant and/or Target Material. Utilisation and Transmutation of Actinides and Long Lived Fission Products. Vienna, Austria, 2003. p. 224

[28] Orlov VV et al. Deterministic Safety of BREST Reactors. In Proceedings of 11-th Int. Conf. on Nucl. Engrg. (ICONE-11). Shinjuku, Tokio, Japan: JSME/ASME; 2003. (CD-ROM file ICONE11-36412)

[29] Orlov VV et al. Fuel cycle of BREST reactors. Solving RW and Non-Proliferation Problems. In Proceedings of 11-th Int. Conf. on Nucl. Engrg. (ICONE-11). Shinjuku, Tokio, Japan: JSME/ASME; 2003. (CD-ROM file)

[30] Alemberti A et al. Proceedings of the V Conf. Innovative Design and Technologies of Nuclear Power. Vol. 1. Moscow: NIKIET; 2018. pp. 20-28

[31] Barinov SV et al. Nuclear safety and fuel efficiency in the reactor BREST-OD-300. In: Proceedings of the XI Seminar on Reactor Physics "Physical Problems of Efficient Use and Safe Handling of Nuclear Fuel"; 4-8 September 2000. Moscow: MEPhI; 2000. pp. 87-90

[32] Gomin EA et al. Status of MCU. In: Advanced Monte Carlo on Radiation Physics, Particle Transport Simulation and Applications. Monte Carlo, 23-26 October 2000. Lisbon, Portugal; 2000. pp. 2003-2004

[33] Emsley J. The Elements. Oxford: Clarendon Press; 1991. p. 255

[34] Scherbak NP et al. Catalog of Isotopic Dates of Rocks of the Ukrainian Shield, Academy of Sciences of the Ukrainian SSR, Institute of Geochemistry and Physics of Minerals, Ministry of Geology of the Ukrainian SSR. Kiev: Naukova Dumka; 1978. pp. 90-136

[35] Okunev VS. Effect of isotopic composition of lead-based coolant of thorium ores on void reactivity effect in BREST reactor. News of higher educational institutions. Nuclear Power. 2006;**2**:56-65

[36] Okunev VS. Reserves of “BREST” concept (at transition to high-capacity power units). In: Scientific Session “MEPhI-2006.” Vol. 8. Moscow: MEPhI; 2006. pp. 89-90

[37] Okunev VS. On the use of thorium ores in nuclear power. In: Proceedings of the 14th Seminar on Reactor Physics “Physical Problems of Fuel Cycles of Nuclear Reactors”; 4-8 September. Moscow: MEPhI; 2006. pp. 111-113

[38] Okunev VS. The impact of different types of MOX-fuel on self-protection BN-type reactors medium and high power. In: Proceedings of Innovate Design and Technologies of Nuclear Power. 3-Ht International Scientific and Technical Conference. 7-10 October 2014. Vol. 1. Moscow: JSC “NIKIET” Publ; 2014. pp. 325-338

[39] Polovinkin VN. Nanotechnologies in power engineering. Moscow: PRoATOM agency; 2010. Available from: http://www.proatom.ru/modules.php?name=News&file=article&sid=2118 [Accessed: 11-01-2010]

[40] Azarenkov NA et al. Nanostructured materials in nuclear energy. Herald of the Kharkov University. Physical Series: Nuclei, Parts, Fields. 2010;**887-1**(45):4-24

[41] Saragadze VV et al. Nanooxidized reactor steels. In: Proceedings of 6th International Ural Seminar “Radiation Physics of Metals and Alloys. Materials for Nuclear and Fusion Power”. Snezhinsk; 2005. p. 78

[42] Okunev VS. Substantiation of the feasibility of using fuel rods with tungsten spraying in power fast reactors of new generation. Thermal Engineering. 2011;**58**(14):1167-1171

[43] Okunev VS. About possibility of further approximation to ideals of natural safety within the concept of the power reactor “BREST” of big capacity. In: Proceedings of Innovate Design and Technologies of Nuclear Power. International Scientific and Technical Conference. 27-29 November 2012. Moscow: JSC “NIKIET” Publ; 2012. pp. 63-74

[44] Rahn FJ et al. A Guide to Nuclear Power Technology. New York: A Wiley-Interscience Publication, John Wiley and Sons; 1984. p. 1000

[45] Kudinov VV. Spraying. Theory, Technology and Equipment. Mechanical Engineering: Moscow; 1992. p. 432

[46] Gurvich LV et al. Energy of Breaking Chemical Bonds. Ionization Potentials and Electron Affinity. Moscow: Nauka; 1974. p. 351

[47] Orlova EA et al. Self-organizing carbonitride coating on molten eutectic lead-magnesium steel. In: Proceedings of Innovate Design and Technologies of Nuclear Power. International Scientific and Technical Conference. 27-29 November 2012. Moscow: JSC “NIKIET” Publ; 2012. pp. 174-182

[48] Shornikov DP et al. Preparation and compaction by plasma-spark and electric pulse sintering methods of nanopowders of uranium nitride. Vector of science of Tomsk State University. 2013;**3**:95-98

[49] Tokita M. Development of advanced spark plasma Sinyering (SPS) systems and its applications. Ceramic Transactions. 2006;**194**:51-60

[50] Okunev VS. Nanotechnology in nuclear power engineering: Pellet cermet MOX-U and MN-U-fuel for fast reactors. In: Proceedings of Innovate Design and Technologies of Nuclear Power. VI International Scientific and Technical Conference. Vol. 1. Moscow: JSC "NIKIET" Publ; 2016. pp. 195-206

[51] Okunev VS. Fine Pellet MOX Fuel for BN Type Reactors. Scientific Session of MEPhI-2015. Vol. 1. Moscow: MEPhI; 2015. p. 130

[52] Wick OJ, editor. Plutonium Handbook. Vol. 2. New-York-London-Paris: GORDON AND BREACH, SCIENCE PUBLISHERS; 1967. p. 446

[53] IAEA-TECDOC-1139. Transient and Accident Analysis of a BN-800 Type LMFR with near Zero Void Effect. Vienna, Austria: IAEA; 2000. p. 243

Chapter 5

Theoretical and Experimental Analysis of Structural Properties of Load-Bearing Components of Thermonuclear Tokamak Installations

Nikolay A. Makhutov, Mikhail M. Gadenin, Sergey V. Maslov, Igor A. Razumovsky and Dmitry O. Reznikov

Abstract

The chapter presents the results of research carried out in Mechanical Engineering Research Institute of the Russian Academy of Sciences that were focused on validation and application of design diagrams, methods and systems for measuring stresses under the modes of Tokamak instillation cooling and management of electromagnetic fields during startups. The examples of tensometric systems and results of measurements of stresses under cryogenic temperatures and strong magnetic fields as well as results of analysis of the states of stresses and strains of structurally heterogeneous components of load-bearing and conductive structures are presented. Operation conditions and limit states of Tokamak components are considered. Results of research summarized in the chapter demonstrate the correctness of the adopted design solutions, which result in a relatively low level of local stresses in the load-bearing components of the thermonuclear installations.

Keywords: thermonuclear installation, strength, service life, local stresses and strains, limit states

1. Statement of problems related to structural integrity and service life of thermonuclear installations

In the modern age, the day-to-day activities of individuals, states and the international community are all closely dependent on the existence of a reliable power supply. The main components of the power industry include energy resources, power installations and power supply systems [1, 2]. Both the global and domestic energy mixes include:

- traditional power sources which have been used for a century or more (thermal - coal, oil, gas, hydraulic, wind, solar);

- new types of energy, which have been used for several decades (nuclear, geothermal, tidal, solar battery, hydrogen, bioenergy);

- promising future energy technologies (thermonuclear, space).

The changes in the respective importance of these types of energy $\overline{E}$ and the relative growth of global energy production $\overline{E}$ in the 21st century are shown in **Figure 1**.

In 2020, Russia adopted a new energy strategy for the period until 2035 [1]. It focuses on the development of new and promising types of energy, with a gradual reduction in dependence on traditional types which is driven by scientific and technological advances and economic efficiency and environmental considerations.

The analysis [1–3] of the prospects for the development of energy until 2030, 2050 and 2100, the focus will be primarily on problems of a scientific, design, technological and operational nature, aimed at ensuring the safety, structural integrity and service life of power facilities [4–8]. This focus is due to the fact that trends in the extraction, production and use of energy have been significantly impacted by major accidents and disasters in the late 20th and early 21st centuries in the USSR, Russia, the USA, Norway, Mexico and China, due to technical failures in unique thermal, hydraulic and nuclear power installations, offshore oil and gas production platforms, and tankers used for the transportation of oil and liquefied natural gas. These accidents and catastrophes were caused by failures to comply with structural integrity requirements and the consequent collapse of load-bearing structures including buildings, vessels, pipelines, electric generators, turbines, and platform structures. Such accidents have resulted in tens or in some cases hundreds of deaths, as well the destruction of technical facilities and chemical pollution and radioactive contamination in surrounding areas. The resulting economic losses are estimated in 10^9–10^{11} US dollars. In this regard, since the 1960s, scientists, engineers and specialists from around the world have begun to pay special attention to ensuring structural integrity and preventing accidents and disasters resulting from technical failures in power facilities of all types. Moreover, fundamentally new tasks have arisen in relation to both existing nuclear energy and promising thermonuclear energy technologies. These types of energy offer new solutions to problems relating to the production, processing and use of energy resources, by virtue of the far smaller volume and mass of fuel required- 1 g of deuterium used in a thermonuclear unit produces the same amount of energy as a whole column of oil tanker wagons. Thermonuclear power, in contrast to nuclear power using thermal and fast

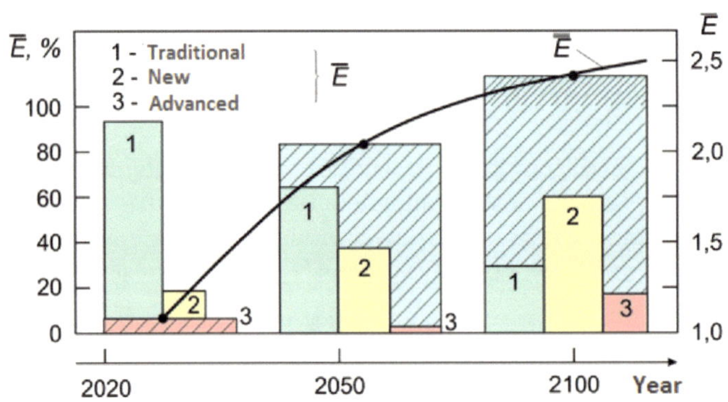

Figure 1.
Structure and development potential of the energy mix.

neutrons, can fundamentally transform the risk of ionizing and radiation damage during the decay and fission of heavy uranium and plutonium nuclei.

The controlled thermonuclear fusion of deuterium-tritium and deuterium-deuterium, which was first proposed in the USSR by academicians A.D. Sakharov, I.E. Tomm and L.A. Artsimovich. This fusion results in the synthesis of heavier helium, accompanied by a release of a huge amount of energy. The thermonuclear fusion processes carried out in thermonuclear power installations on Earth are essentially the same as the processes that take place on the surface of and inside the Sun.

The subsequent theoretical research on, and practical development [2] of thermonuclear fusion in Russia was led by Academicians E.P. Velikhov, V.A. Glukhikh, and B.B. Kadomtsev, who worked on the physics of thermonuclear installations, and Academician K.V. Frolov [2]., who worked on issues relating to mechanics and the structural integrity and service life of facilities. The initial reactors developed for use in thermonuclear installations were of two types:

- pulsed reactors in which small targets with a deuterium-tritium (D-T ^2H,^3H) mixture are heated within short-term cycles (τ = 10^{-8} sec) by a dynamic powerful electron and laser beams so as to trigger a micro-nuclear explosion at ultrahigh pressures;

- quasi-stationary systems (Tokamaks), in which such mixtures are heated to a plasma state and held for periods τ <1 sec by strong magnetic fields at low pressure and ultrahigh temperature $t > 10^8 \div 10^9$°C.

The resolution of the problems relating to thermonuclear power installations can be (**Figure 1**) divided into three main stages:

- the creation and use of research reactors (1960–2000);

- the creation of demonstration reactors with a positive energy yield (2000–2030);

- the anticipated creation of industrial reactors producing energy on an industrial scale in thermonuclear power plants (2050–2100).

In Russia, the first stage included a pulsed thermonuclear installation of the first type, Angara-5 (**Figure 2**) in which a micro-fusion explosion was triggered by a stream of electrons emitted from a system of super-powerful condensers, and a series of fusion reactor installations of the second, or tokamak type (T-7, T-10, T-15, T-20).

The emergence and behavior of thermonuclear fusion of a deuterium-tritium mixture with the formation of helium occurs in Tokomak installations, with the release of high energy. Maintaining this reaction in the toroidal chamber is carried out by a powerful alternating magnetic field generated by the network of coils. The thermonuclear fusion reaction is carried out in the plasma inside the chamber, heated to a temperature of over 100 million degrees. Huge mechanical and thermal forces arise in electromagnetic coils with a superconductor at cryogenic temperatures. These forces generate high stresses and strains in Tokamak structures (**Figure 3**).

Drawing on research into all types of reactors [4–10] on the basis of the above developments of experimental thermonuclear installations, in 1995 an international work (USA, USSR-Russia, Japan, India, France, South Korea and others) began on the development [3, 11, 12] of the world's largest international demonstration thermonuclear

Figure 2.
Angara-5.

Figure 3.
Tokamak T-7.

Figure 4.
International thermonuclear experimental reactor ITER (ITER).

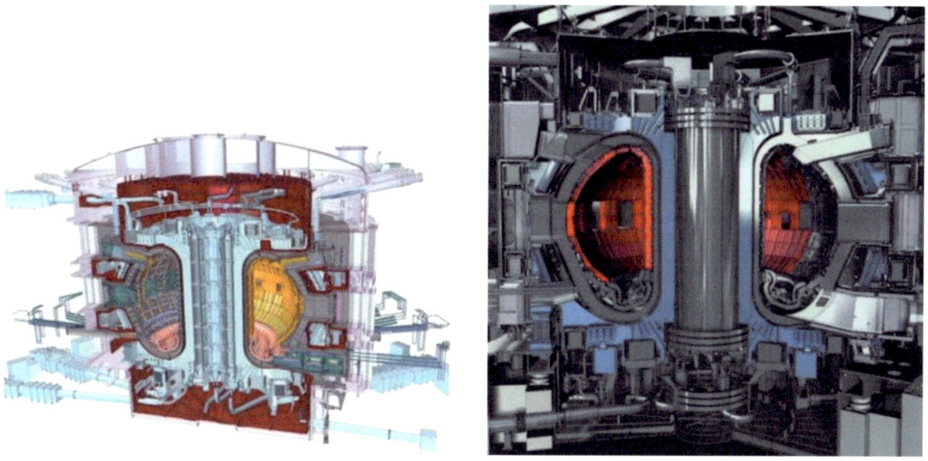

Figure 5.
Schematic diagram of the international thermonuclear experimental reactor ITER.

reactor, or ITER (**Figure 4**) with a chamber volume of 830 m^3 and a plasma temperature of 100–500 million ^{0}C (**Figure 4**), to be completed by 2025 (**Figure 5**).

2. Analysis of special operating conditions and limit states

Experimental thermonuclear reactors of the tokamak type are fundamentally different from the nuclear power plants (NPPs) using tested thermal and fast neutron [4] technology that is currently in operation. These specific features of the experimental reactors include:

- a wide range of low operating temperatures, *t, f*rom room temperature of $t = +20^{0}$C to cryogenic temperature of -269^{0}C (liquid helium temperature) in superconducting systems of electromagnetic coils for the creation of magnetic fields;

- high temperatures (up to +800÷1000^{0}C) on the working walls of the chamber and ultra-high temperatures (up to +100000000^{0}C) in the plasma;
- ultra-high magnetic fields M for holding and controlling plasma from 3 to 20 Tesla (T);
- controlled electromagnetic fields of unstable duration τ, with a frequency of change f of up to 1000 Hz;
- extremely high mechanical $Q_M(\tau)$, thermal $Q_t(\tau)$ and $Q_{em}(\tau)$, electromagnetic forces $Q_{em}(\tau)$ reaching 100 ÷ 1000 MN;
- ultra-high density, i,of the electric current in the coil superconductors;
- widely varying mechanical and physical properties (coefficient of thermal expansion α), elasticity modules E, yield stress σ_y, and ultimate strength σ_u of the conjugated structural composite materials (superconducting wires, conductors, coils, support structures, multilayer chamber walls);

The design features and loading conditions, listed by the parameters: τ, Q_M, Q_t, $Q_{em}(\tau)$ t, E, σ_y, σ_u σ_u , α are combined with standard and emergency situations (for example, loss of superconductivity and contact of plasma with the chamber wall).

When justifying the strength of load-bearing structures, the two most important tasks are:

- to determine the stress-strain states in hazardous areas during normal and emergency situations;
- to determine the limit states resulting in high levels of plastic deformation and fracture.

The limit states in the load-bearing structures of the tokamak will be related to:

- the achievement of critical fracture stresses σ_c and strains e_c under extreme loads $Q_{max}(\tau)$;
- the achievement of maximum permissible strains, $e_{k,}$ in superconductors at the stage of loss of superconductivity when the reactor is cooled down to critical helium temperatures and when the maximum electric current I_{max} with a density i_{max} is introduced into the superconducting coils;
- the occurrence of critical damage as a result of exposure to a combination of variable mechanical $Q_m(\tau)$, electromagnetic $Q_{em}(\tau)$, thermal $Q_t(\tau)$ and contact $Q_k(\tau)$ forces.

The above are the basic parameters for determining the local stresses, σ, the strains e, strength and service life of a thermonuclear installation [6].

$$\{\sigma, e\} = F\{Q(\tau), E, \mu, \tau, t, S\} \tag{1}$$

$$\sigma_{\max} \leq \left\{\frac{\sigma_y}{n_y}, \frac{\sigma_u}{n_u}\right\} \tag{2}$$

$$\tau \leq \frac{\tau_c}{n_\tau}, N \leq \frac{N_c}{n_N} \tag{3}$$

where F_σ is functional, S is the characteristic of the dangerous section, σ_{max} is the maximum stress at the critical point of the dangerous section, n_y, n_u are safety factors for yield stress and ultimate strength, τ_c, N_c is the critical time until fracture (service life) and n_τ, n_N safety factors for service life.

For thermonuclear installations $n_y < n_u < n_\tau \approx n_N$.

The creation of energy-efficient and safe thermonuclear tokamak installations is largely dependent on the successful solution of the problems of deformation and fracture mechanics.

A characteristic feature of large tokamaks as mechanical systems is the presence of significant ponderomotive loads $Q_{em}(\tau)$, which act in combination with the special operating conditions in the main systems of the installation - the electromagnetic system (EMS) for plasma confinement and the discharge chamber.

The electromagnetic systems used in tokamaks give rise to high and ultrahigh forces $Q(\tau)$, which results in high mechanical stress on the structural elements. Moreover, in superconducting EMS, structural, current-carrying and insulating materials operate at cryogenic temperatures (down to 4.2 K), which affect the physical and mechanical properties of these materials. The discharge chambers of tokamaks are exposed to complex mechanical, thermal, and radiation loads.

One specific feature of tokamaks is that they operate in an alternating programmed stable and unstable mode, and so their power elements are subject to cyclic loads. The calculation and design of such elements needs to be carried out based on their actual operational characteristics, using an apparatus with static, dynamic and cyclic strength and taking into account plastic deformation of the materials.

The program of research into controlled thermonuclear fusion technology provides for an increase in the size and intensity of magnetic fields and more complex operating conditions of tokamak coils. Moreover, the standards in relation to the durability, rigidity, and reliability of load-bearing elements are also increasing, which will ensure that the required physical parameters are complied with in respect of the facilities. There is therefore a pressing need to consolidate the expertise in solving problems related to mechanics, durability, service life and safety issues which has been accumulated during the development of the tokamaks used in the largest thermonuclear installations. In Russia, these were the T-15 tokamak and the Strong Field Tokamak (SFT) installations.

3. Experimental study of stress-strain states

Computational and experimental studies of the deformation fields of a structure make it possible to determine the relationship between the levels of stress-strain fields in hazardous zones of the structure which are inaccessible for direct experimental research, and in areas which are accessible for the purpose of continuous observation. The determination of these functional dependencies forms the experimental and theoretical basis for monitoring the operational performance of the installations.

The main stages of research into the stresses to which the critical elements of power installations are exposed include:

- an analysis of the initial data on -magnetic fields, ponderomotive loads and boundary conditions which is necessary for research using models, full-scale elements and mathematical calculations;

- the development of algorithms and programs to perform mathematical calculations and process of experimental results;
- the theoretical optimization of the design parameters;
- the development of methods and instruments to measure deformations in cryogenic conditions and strong magnetic fields;
- the development of systems for full-scale measurement of strain at the main nodes of the installations;
- testing of materials to determine basic design characteristics;
- research on various types of model - photo-elastic, low-modulus, and models using strain-sensitive coatings;
- full-scale inspection of the main components of installations in operational conditions and near-operational conditions, for strain measurement purposes;
- analysis of the results in order to assess strength and service life of the structure.

Below are some of the main results of the research on the T-15 installations.

The T-15 installation is designed to create and study plasma with parameters that are close to the thermonuclear level and are sufficient for a reliable transition to the plasma state. One of the main technical features of the T-15 installation is the use of superconducting toroidal field coil (STFC).

Each coil unit contains coils of a superconducting current-carrying components, enclosed in a rigid steel case. The coil is a transversally isotropic ring made of a complex composite material consisting of a niobium-tin superconductor in a copper matrix, insulating materials and channels for a coolant (helium).

The T-15 unit contains 24 STFC units located around the central support cylinder along the torus-shaped vacuum chamber. A structural diagram of the SCTF unit is shown in **Figure 6**.

It has been established that as a result of the interaction of the STFC currents with the toroidal and poloidal magnetic fields in the STFC units, two types of volumetric ponderomotive loads arise:

- toroidal static forces acting in the plane of the STFC unit. The resultant force of this load is directed towards the center of the installation;
- poloidal impulse forces acting on the coil perpendicular to the its plane. These lateral forces create a tilting moment relative to horizontal diameter of the coil.

For the purposes of the operation of the T-15 unit, provision is made for nominal and forced modes with and without disruption by plasma current. The values of the loads acting on one unit are shown in **Table 1**.

The maximum strain of a superconductor in any mode may not exceed 0.2%, since the current-carrying capacity of superconducting systems (SCS) drops sharply at strains of more than 0.5%.

Due to the symmetry of structure and loads, all STFC units are subject to the same conditions. Therefore the calculation and experimental study of their

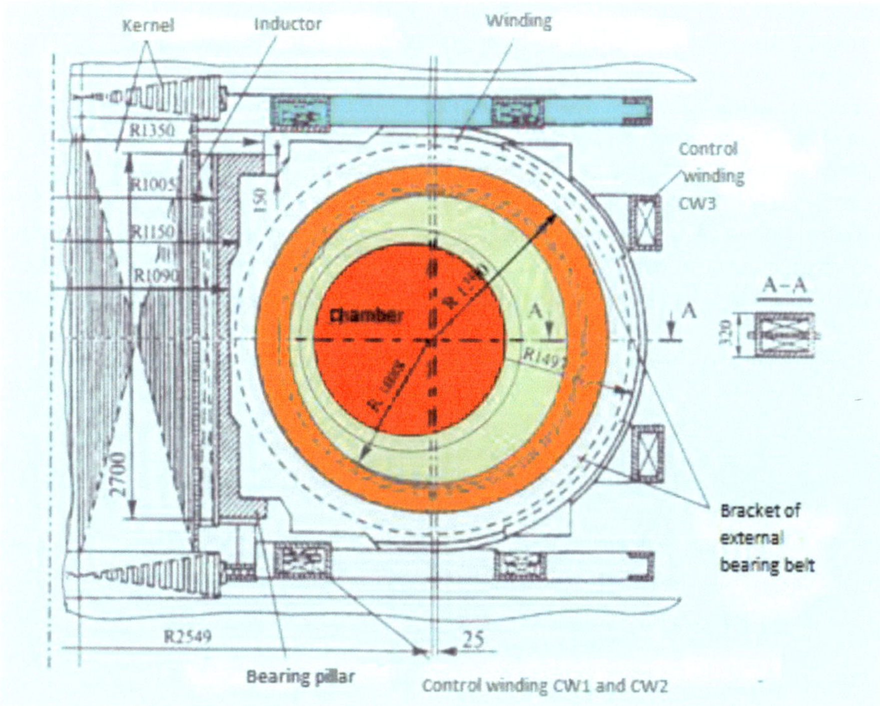

Figure 6.
Structural diagram of the STFC unit.

T-15 operation mode	Central force, MN	Overturning moment, MN·m	Number of loading cycles
Nominal:			
without disruption	6.07	1.45	80,000
with disruption	6.07	2.4	400
Forced:			
without disruption	12.4	2.17	20,000
with disruption	12.4	3.6	100

Table 1.
Loads acting on the unit.

stress-strain state (SSS) is carried out taking into account the corresponding symmetry and support cylinder conditions.

To study the effects $Q(\tau)$ of stresses $\sigma(\tau)$ and strains $e(\tau)$ associated with the action of magnetic and thermal fields, foil strain gauges with sensitive elements made of constantan and wire (wire diameter 30 μm) and high-temperature tensoresistors made of a nickel-molybdenum and iron-chromoaluminum alloy were used. The tensoresistors were installed on samples consisting of three materials, simulating the main structural materials of the tokamak: stainless steel, copper alloy and a composite material. As a result of the first series of experiments, it was found that in magnetic fields of 1–3 T the response of the output signals from the strain gauges was $20x10^{-6,}$ with standard deviations of about $S = 1.5 \times 10^{-6,}$.

To develop tensoresistors with optimal characteristics at cryogenic temperatures, nickel-molybdenum tensoresistor alloys with a low electrical resistance coefficient in the temperature range of -269 to = 20^0C (4–300 *K*) were created. Under appropriate heat treatment modes (**Figure** 7) these alloys have a high residual electrical resistance K_t at ultralow temperatures down to $0.5 \times 10^{-6} K^{-1}$.

Figure 8 shows the layout of the primary converters and the results of the study of stresses during cool-down and the injection of current into the system consisting of two experimental STFC units located side by side and not aligned. With the units arranged in this way, the interaction of their fields creates a load that approximately corresponds to the load on the units in the operating mode. It has been established by measurement that at a cooling rate of up to 3 K/h, the maximum stresses arise at temperatures up to 60 K and do not exceed the yield stress σ_y. With a further decrease in temperature, the level of the stresses decreases, and

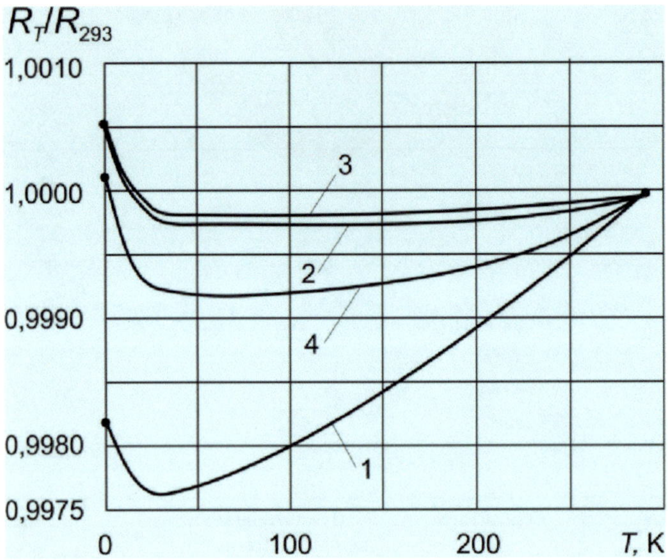

Figure 7.
Temperature correlation between electrical resistance after quenching and various stabilizing annealing modes (1 - quenching, 2 - quenching and annealing for 30 minutes at a temperature of 470°C, 3 - annealing for 2 hours, 4 - annealing for 5 hours).

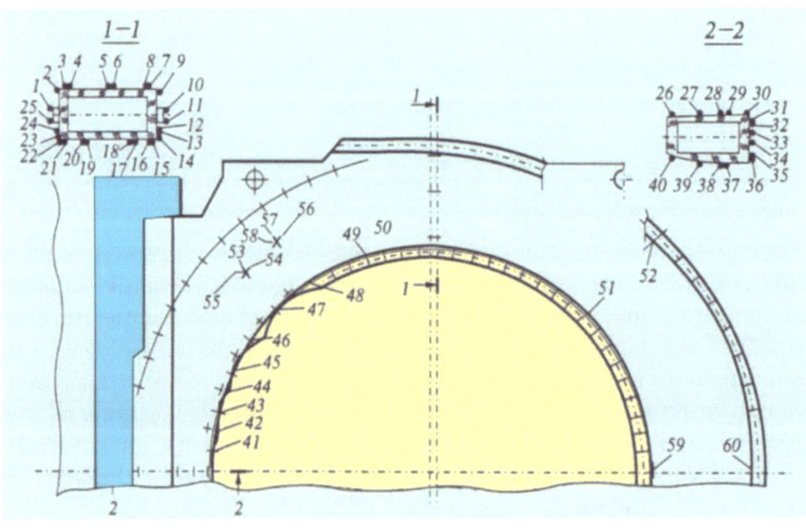

Figure 8.
Arrangement of tensoresistors on the strain gauge model of the case of the STFC unit in the T-15 installation (the figures refer to the - numbers of the tensoresistors located along the diameter, across and at an angle of 45°).

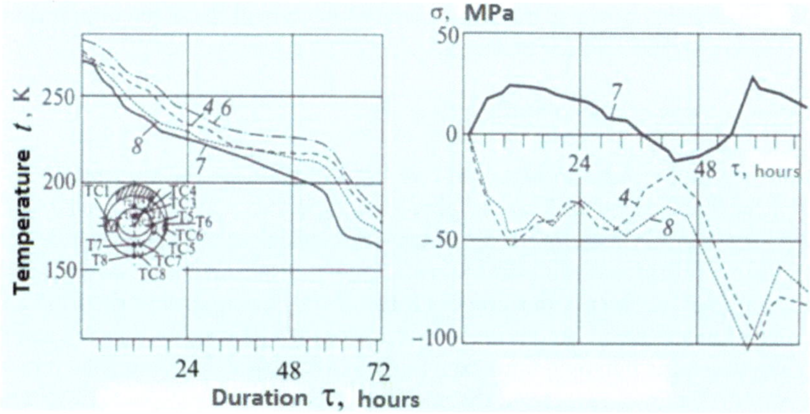

Figure 9.
Temperature and stresses in the housing of the STFC unit during cool-down.

the temperature field becomes more uniform. The measurement of displacements of the SCS relative to the body during the initiation of the current revealed shifts of up to 15 mm, which could lead to the delamination of the SCS from the walls of the body. On the basis of these measurements, measures were taken to increase the stiffness of the sealing of the coils in the body of the standard STFC units to ensure their operational fitness.

The results of the temperature $t(\tau)$ and stress measurements $\sigma(\tau)$ during cooling are shown in **Figure 9**. The layout of the resistance thermometers, RT, and tensoresistors, T, is shown under the curves.

The numbers of the curves correspond to the numbers of resistance thermometers RT.

Measurement of the changes in stresses when current - from 0 to 15 kA - was introduced into the superconducting systems showed that in the support cylinder the stress reached 110 MPa, and 40 MPa when the discharge chamber was heated.

4. Calculations and physical modeling

The initial computational study of the stress-strain state of the STFC unit as a result of the action of toroidal forces in a complete setting provides a solution to the spatial problem. By using equivalent elasticity modules and taking into account the nature of the load, this problem can restated more simply, in two dimensions. The nature of the SSS in radial sections of the STFC, which are remote from the support column, can be investigated in an axisymmetric setting. The SCS coil housed in the steel power case of the STFC is anisotropic in the circumferential and radial directions. These problems can be solved using the finite element method.

At the initial design stage, calculations were carried out in order to select the best design option. Then, for the selected design option, a study of the stress-strain state of the STFC unit was carried out in relation to the refined design schemes using the finite element method and physical modeling, which allows specific features of the design to be taken more fully into account.

The influence of the following factors was studied:

- the bending stiffness of the support cylinder;
- the radial technological gap between the STFC units and the support cylinder;

- the volumetric nature of the application of loads and the force interaction between the case and the SCS coil;

- the anisotropy of the coil properties.

The superconducting toroidal field coil (STFC) has a strong anisotropy in respect of its mechanical and thermophysical properties. On the planes adjacent to the strips separating the half-shells, contact friction arises during each pulse, which in cryogenic conditions is undesirable from the point of view of heat release and insulation integrity. The ponderomotive forces which compressing the STFC in the radial direction and stretching it in the circumferential direction during each pulse cause gaps to appear between it and the body bandaging it. The significant lack of uniformity in the mechanical and thermophysical properties of the SCS causes a significant lack of uniformity in the stresses to which it is subject, which can lead to appearance of plastic deformations and accumulation of residual stresses in the SCS.

The modeling of strains and stresses in the SCS was carried out using polarization and optical methods.

The calculations of the SSS of the STFC unit, taking into account all the above factors, are shown in **Figure 10**.

In order to ensure that the electrical insulation is reliable, the contact interaction at the node where the support column is connected with a pin to the metal-polymer coils of the STFC (see **Figure 6**) needs to be calculated and assessed. The stress state of the node is almost flat and skew-symmetric. The initial contact takes place near the corner point (**Figure 11a**). At this point of contact, the stresses are very high, which may lead to destruction of the polymer coating. The simplest stress-limiting change in a contact surface is the - rounding of a sharp edge. However, the stress distribution remains significantly uneven (**Figure 11c**), with a sharp increase in stresses near the rounded edge at point A. Analysis of similar options for the contact interaction between a rigid punch and an elastic layer shows that a more favorable pressure distribution takes place when the punch is convex in shape, in which case the curve on the diagram is close to parabolic, with zero pressures at the boundary of the contact area (**Figure 11c**).

The determination of the SSS of the STFC unit resulting from the action of poloidal forces, Q,- is a complex spatial problem involving the mechanics of a

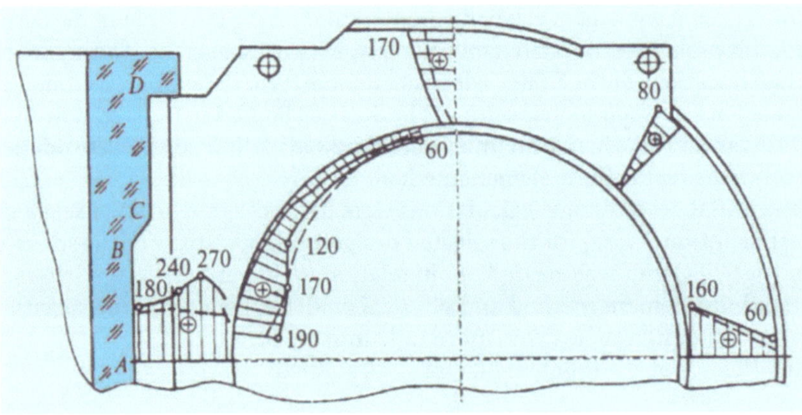

Figure 10.
Circumferential stresses in the unit body when it is supported by the column along the AB line; solid lines (-) - values reached by experiment; dotted lines (--) - calculated values.

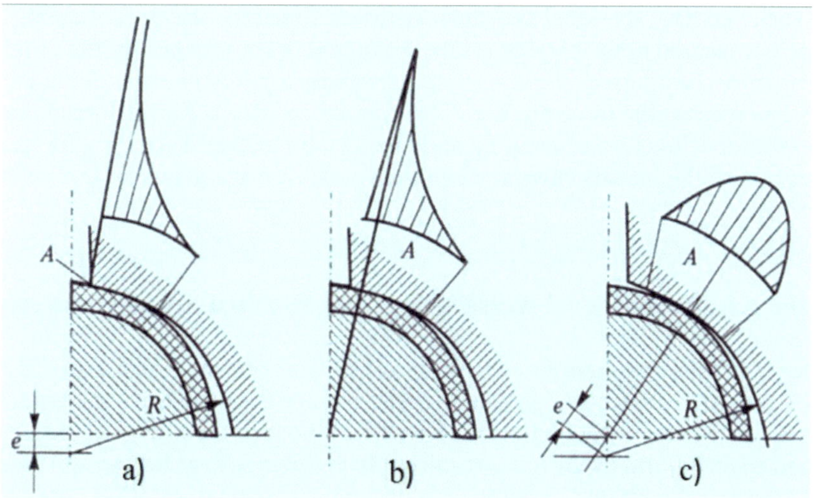

Figure 11.
Distributions of contact pressures on a metal-polymer pin with various hole shapes and a split support column with a sharp (a) and rounded (b) edge and (c) with a displacement of the point of initial contact A.

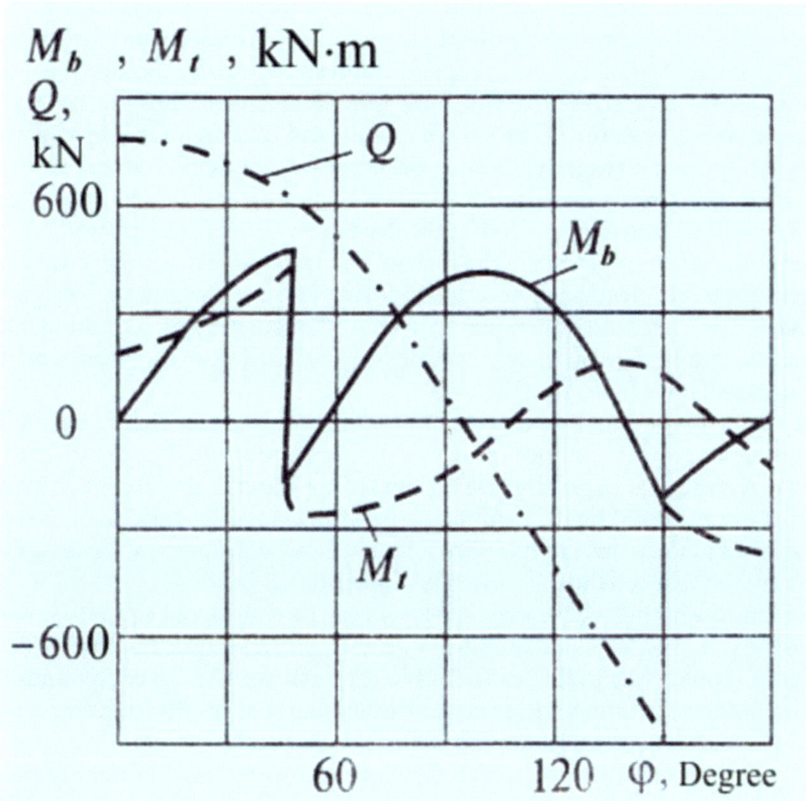

Figure 12.
Distributions of bending and torque moments and shearing forces arising in the forced mode when the plasma current is disrupted.

deformable solid body. This task was solved using a pivotal approach. To determine the SSS, the rod theory is used: this theory takes into account the potential energy of bending, torsion and antiplane shear deformations. A rod is viewed as an elastic curve with bending, torsional and shear stiffness. Since the stiffness of an SCS coil with such indicated deformations is low, compared to the stiffness of the unit body, the former can be ignored, thus increasing the design safety margin. The calculation is performed using the force method. The calculation results for the forced mode, in the form of distributions of bending and torque moments and shearing forces, with a disruption of the plasma current when these loads are at a maximum, are shown in **Figure 12**.

5. Design analysis of local stresses in composite structures of the reactor

The intra-chamber components of thermonuclear reactors, including ITER [3, 6, 11, 12], are subject to high stress levels, and are the most critical elements as they are in direct contact with the plasma. Taking into account p. 2, they are designed to withstand cyclic loads resulting from intense heat flows and volume forces, thermal shocks and dynamic effects during plasma disruptions and abrupt displacements of magnetic axes. The operating conditions to which the materials used are subjected are complicated by exposure to radiation.

When designing intra-chamber components, it is very difficult to find a structural material that is sufficiently resistant to all the above factors simultaneously and can provide the structure with the required operability and service life. The intra-chamber components were therefore constructed using layers of materials with different qualities: beryllium, copper, stainless and austenitic steel.

In view of the high stresses to which the individual layers whose surface is directly in contact with the plasma, are exposed, and their increased brittleness, it is of great importance to apply fracture mechanics methods in order to confirm and ensure their integrity.

The main load factor to which the multilayer elements of the first wall are exposed is the effect of plasma in the form of cyclically repeating high power heat flows and changing electromagnetic loads during plasma disruptions. Design strength analysis needs to be conducted in respect of such elements, taking into account possible changes in their properties during heating and increased brittleness due to radiation [4–8, 11, 13].

In addition to thermal loads caused by varying temperature fields, areas adjacent to the boundaries of the composite layers experience additional loads due to the difference in the thermal expansion coefficients of the dissimilar materials. Moreover, residual stresses caused by the manufacturing process are localized in this area. Due to differences in the materials' physical and mechanical properties, these residual stresses are not relieved during subsequent operational heating.

In effect, the boundary between the layers can be considered as another material, with own initial level of damage and defects, and a specific fracture resistance. For example, one of the problems with obtaining efficient Be/Cu compounds was that combining beryllium with almost any other material results in the formation of brittle intermetallic phases [5, 14, 15].

The operating conditions to which the dissimilar materials used for the structure are subjected are complicated by exposure to radiation, which increases the likelihood of fracture due to brittleness. The fracture resistances of the materials used (beryllium, copper and stainless steel), taking into account increased brittleness due to radiation, are set out in the report [16].

To maintain the proper operational condition of the multilayer element which is in contact with the plasma, it is necessary to ensure both the integrity of the reinforcing material (beryllium) and the proper operational condition of the beryllium-copper joint. Possible damage (failure) scenarios during operation include the initiation of a brittle fracture in the beryllium layer, and delamination at the beryllium-copper boundary.

For the purpose of assessing the damage resistance of a heterogeneous structure, it can be divided into three separate zones: areas remote from the boundary between layers of dissimilar materials; areas near the boundary; and the boundary itself.

To assess the fracture resistance of layers remote from the boundary, fracture mechanics appropriate to a homogeneous material can be applied. The applicable theoretical and practical foundations applicable to this case have been developed [17, 18].

To analyze the resistance to brittle fracture of the materials in the layers located near the boundaries, the methodology appropriate for a homogeneous material is applied, but the corresponding analytical formulae for calculating the criterion parameters need to be adjusted [19]. This applies to the boundary zone where dissimilar materials are joined together, which is, at present, the area in respect of which least research has been done.

The main factors to be considered are as follows.

- The boundary between dissimilar materials forming a junction are sources of stress and strain singularities. This applies both to cracks occurring at the boundaries between dissimilar materials (delamination cracks), and to the points where the boundaries between materials exit to a free surface.

- In order to calculate the stress-strain state (SSS) in such singular zones using modern mathematical methods, special finite elements need to be developed whose shape functions allow the features of the SSS in the zones of crack tips or corner points, as well as in their immediate vicinity, to be taken into account [20, 21].

- The materials used in multilayer elements do not have levels of fracture resistance: rates: copper is prone to radiation embrittlement and beryllium is a fragile material. The durability and fracture resistance characteristics of laminated structures largely depend on the technology used in their manufacture.

In view of the above, in accordance with currently accepted design practice the calculation of the durability of multilayer elements in reactors needs to be carried out in two stages.

1. The first stage is the assessment of strength of multilayer elements without taking into account the singular zones and possible fractures, which can be performed on the basis of classic criterion-based approaches, taking into account the known characteristics of the materials used in a multilayer composition. In doing this, modern methods of mathematical modeling need to be used to analyze the SSS under power and temperature loads.

2. The second stage is the calculation of structural elements using fracture mechanics methods, taking into account the special characteristics of the SSS in singular zones (strains) and postulated crack-like defects, in order to assess safety coefficients while taking brittle fracture into account.

In the case (**Figure 13**) of a homogeneous plate with a fracture l under nominal loads σ [5–8, 17–20, 22] the distribution of local stresses σ_r at a distance r from the crack tip is described by a singular equation:

$$\sigma_r = \frac{K_I}{\sqrt{\pi r}} \cdot f_\kappa = \left(\frac{K_I}{\sqrt{\pi}} \cdot f_\kappa\right) r^{\lambda} \tag{4}$$

where K_I is stress intensity factor of at the crack tip ($K_I = \sqrt{\pi l}$), f_κ is a dimensionless function, which depends on the dimensions of the plate and crack, and λ is the singularity index ($\gamma = -0.5$).

In accordance with the strength conditions (1) and (2) for a plate without a fracture ($l = 0$) and where a fracture is present as shown in **Figure 12**, the fracture resistance condition is written in the form

$$K_I \leq \frac{K_{Ic}}{n_\kappa} \tag{5}$$

where K_{Ic} is the critical stress intensity factor and n_κ is safety factor for fracture resistance ($n_\kappa \leq n_u$).

During the analysis of the design features of the multilayer elements of the first wall of the thermonuclear reactor [14, 23, 24] several main types of singularity sources were identified (**Figure 14**):

- singularities resulting from including in the calculation postulated defects (fractures) located in a homogeneous layer of a multilayer composition (**Figure 14a**);
- a delamination fracture located at the boundary between dissimilar materials (**Figure 14b**);
- a fracture adjacent to the joint boundary (**Figure 14c**);
- joints of dissimilar materials: beryllium-bronze 90° -90° and 90°-180° (**Figure 14d**, **e**).

In a situation as per scheme 14a, in which a fracture (discontinuity) is completely located in one of the homogeneous materials of a multilayer element,

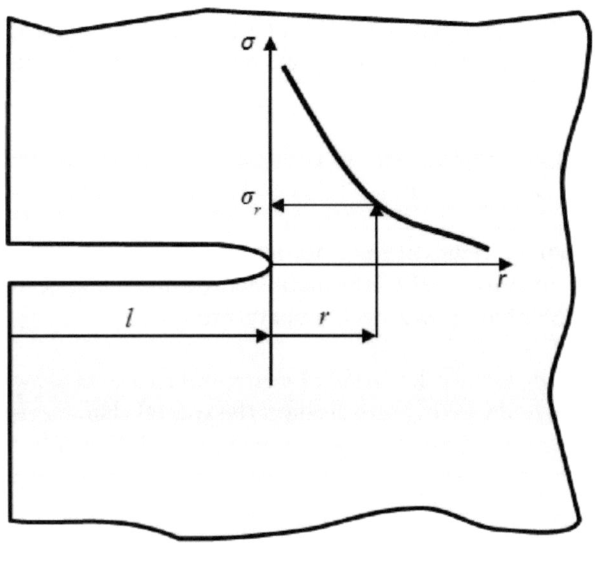

Figure 13.
Calculation method for analyzing the distribution of stresses σ_r in the crack zone in conditions of plate tension.

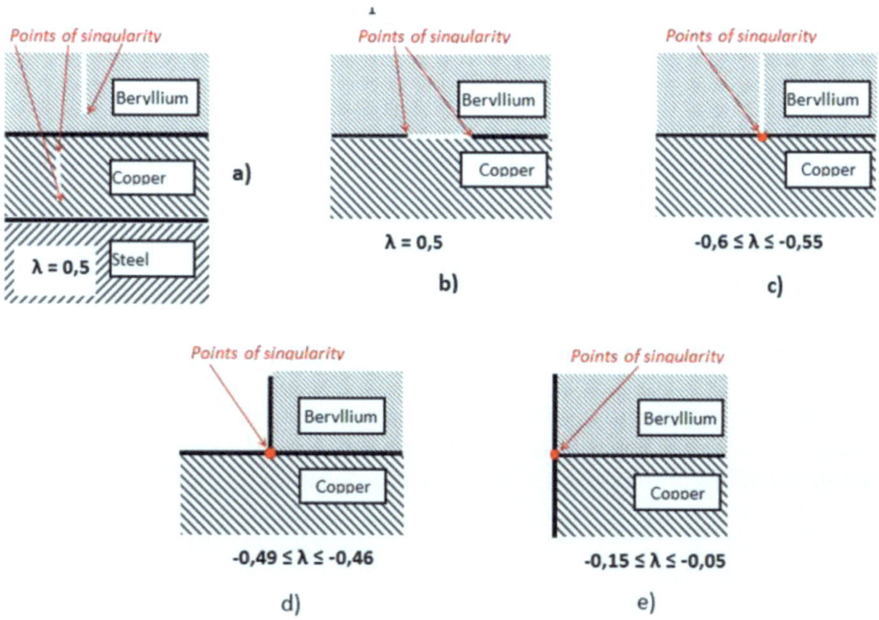

Figure 14.
Basic types of sources of stress singularities in multi-layer elements of the first wall of reactor. (a) defects (cracks), located in homogeneous material; (b) delamination crack, located at the boundary of heterogeneous materials; (c) crack, abutting on the boundary of a joint; (d) joints of heterogeneous materials Beryllium-Copper 90^0–90^0; (e) joints of heterogeneous materials Beryllium- Copper 90^0–180^0.

the procedures for calculating the SSS and durability are well developed ([17, 18, 22]). The level of stress singularity indicator in this case is $\lambda = -0.5$.

For delamination fractures (**Figure 14b**), subject to loads by the mechanisms of normal separation and transverse shear, the characteristic Eq. (4), corresponding to the solution of the characteristic equation that determines the degree of stress singularity - λ has a complex root [17, 19, 25].

$$\lambda = -1/2 \pm \beta;\ \beta = \frac{1}{2\pi}\ln\left(\frac{\mu_1 + \mu_2\kappa_1}{\mu_2 + \mu_1\kappa_2}\right) \tag{6}$$

where μ is the shear modulus of the materials, κ = 3-4 ν (for plane deformation), ν is Poisson's ratio and the indices 1 and 2 indicate whether the material is the 1st or 2nd in the compound of dissimilar materials.

- deffects (cracks), located in homogeneous material (**Figure 14a**);
- delamination crack, located at the boundary of heterogeneous materials (**Figure 14b**);
- crack, abutting on the boundary of a joint (**Figure 14c**);
- joints of heterogeneous materials Beryllium- Copper 90^0–90^0 и 90^0–180^0 (**Figure 14c, d**).

From (4)–(6) it follows that the asymptotic distribution of stresses at the fracture tip at $r \to 0$, $\beta \neq 0$ is singular, with a different λ singularity. In a bulk (three-component) stress state, three models of fracture mechanics (I, II, III) are introduced into the calculation [18–20], and then, based on (4)

$$\sigma_r = \frac{K_I}{\sqrt{\pi}} \cdot f_{KI} r^{\lambda_I} + \frac{K_{II}}{\sqrt{\pi}} \cdot f_{KII} r^{\lambda_{II}} + \frac{K_{III}}{\sqrt{\pi}} \cdot f_{KIII} r^{\lambda_{III}} \quad (7)$$

The proposed approach to assessing strength of adapters will require additional study of residual technological stresses, which, due to the difference between the physical and mechanical properties of adapter materials, always reach significant magnitudes.

6. Strength and durability analysis

Measurement of the changes in stresses $\Delta\sigma$ when electric current - from 0 to 15 kA - was introduced into the superconducting systems showed that in the support cylinder the stress reached 110 MPa, and 40 MPa when the discharge chamber was heated. During the experimental testing and operation, the elements of the discharge chambers of tokamaks are exposed to mechanical $Q_m(\tau)$, temperature $Q_t(\tau)$ and electromechanical $Q_{em}(\tau)$ loads, some of which cause the presence of repeated elastoplastic deformations in the areas of stress concentration. As a result, it was necessary to study and substantiate the static and cyclic strength, for which a series of experimental studies was carried out involving the single-frequency and dual-frequency loading of austenitic chromium-nickel stainless steel at a wide temperature range t from −196°C up to 400°C. The resulting characteristics of the material's resistance to static and cyclic deformation and destruction are an integral part of the general design justification in relation to the strength and durability of the N elements of the discharge chamber.

The calculations and tests were carried out in relation to the tokamak installation in the presence of a strong magnetic field, the purpose of which was to conduct a physical experiment to study the behavior of plasma under conditions close to those of thermonuclear ignition at minimal technical and economic costs. The use of strong magnetic fields to confine plasma makes it possible to significantly reduce the size of the electromagnetic system of the tokamak and the amount of energy stored in it, and the use of combined adiabatic compression - significantly increases the potential of the experiment. However, in order to increase the magnetic field, a number of complex engineering problems need to be solved.

Structurally, the SFT consists of a discharge chamber in the form of a closed torus of noncircular cross-section, along which 32 sections of the toroidal field coil (TFC) are located. The poloidal field coils (PFC) are located outside the TFC. A structural diagram of the electromagnetic system (EMS) is shown in **Figure 15**.

The interaction of TFC currents with toroidal and poloidal fields leads (by analogy with **Figure 12**) to the occurrence of significant ponderomotive forces acting in the TFC plane and overturning moments tending to rotate the TFC section planes around their horizontal axes. The total vertical force, Q, disrupting the TFC, is equal to 128 MN, the resulting centripetal force is 108 MN, and the magnitude of torque relative to the vertical axis of the installation is 30 MN·m. In addition to the forces caused by the interaction of magnetic fields and currents, significant forces arise in the TFC, which are a consequence of heating of the conductor.

Ensuring strength of the TFC under the influence of these forces is one of the most difficult tasks involved in developing an EMF. Each section of the TFC is made in the form of a single-turn conductor (bus) made of zirconium bronze, placed in a band made of high-strength non-magnetic steel. The TFC sections are interconnected in such a way that, as a result, a closed thick-walled toroidal shell is formed. This structure is capable of handling both azimuthal and centripetal loads.

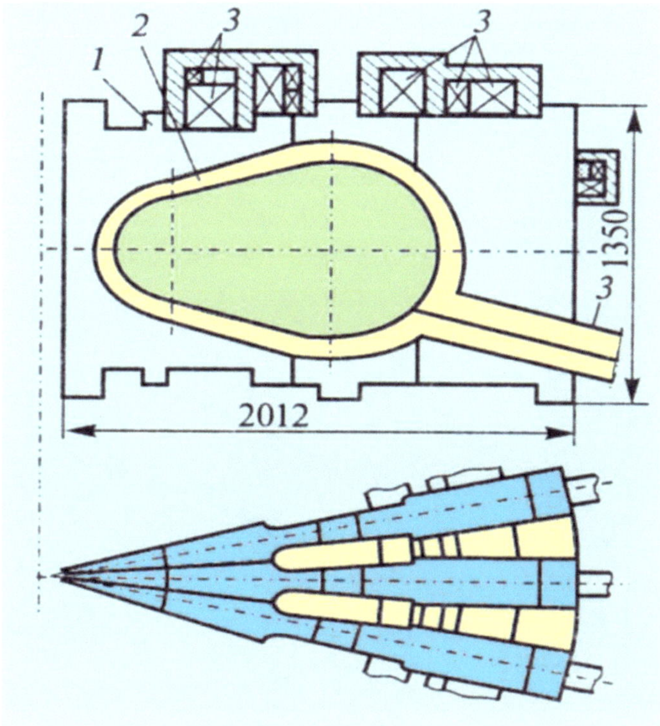

Figure 15.
Structural diagram of the SFT electromagnetic system. 1 - bandage; 2 - bus; 3 - poloidal field coil.

It has been established by analysis that it is not possible to solve the problem of ensuring EMS strength by means of traditional safety factors and acceptable stresses. The development of the structure was therefore carried on the basis of the maximum strength characteristics of the material. This approach is acceptable, since the unit is designed for a limited number of impulses.

The selection of the structure of the section, bandage and other load-bearing elements was made by considering the stress-strain state of a number of design options, taking into account the elastoplastic behavior of materials under the action of ponderomotive forces and thermal stresses. The problem of studying the stress-strain state of a structure while a conductor is undergoing elastoplastic deformation can be solved using the finite element method, by applying the theory of plastic flow. Analysis of the stress state has shown that the effect of overturning moments on maximum stresses and strains is insignificant. **Figure 16** shows the distribution of maximum stresses in the TFC unit resulting from the action of ponderomotive forces on the coil in the unit plane, taking into account the heating of the conductor. An assessment of the service life of the EMS has shown that it is able to withstand a given number (1000) of full-scale operating cycles.

The determination of pulse loads on the inner surface of the magnet coil based on the measurements of stresses arising from such loads in individual zones of the structure is an inverse problem of experimental mechanics.

The resolving equation connecting the stresses determined from measurements in a certain zone (**Figure 17a**) with the required load vector on a part of the surface is expressed in the form of a system of Fredholm integral equations of the first kind. It is pointless to attempt to resolve this system - this would be an incorrectly posed problem, —as small perturbations of the initial data can result in arbitrarily large perturbations of the solution. A regularizing algorithm is therefore chosen. The

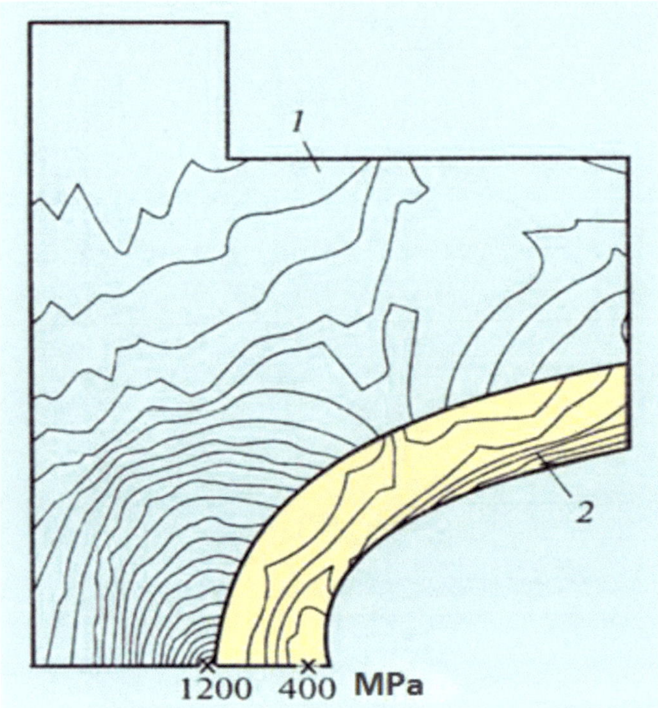

Figure 16.
Isolines of equivalent stresses acting in the bandage (1) and the bus (2). The asterisks mark the points of local maxima.

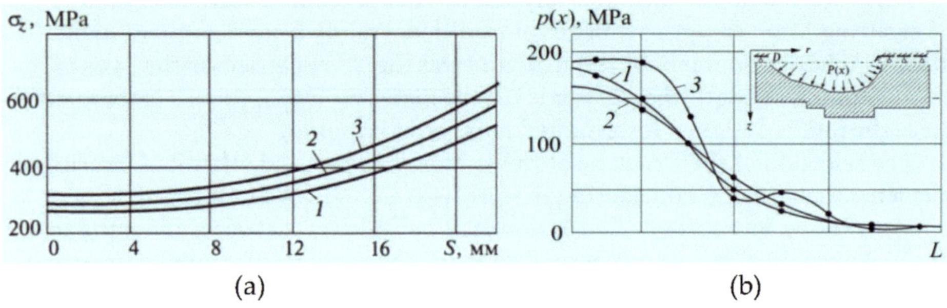

Figure 17.
The principle stresses in a narrow section of the TFC bandage (a) and the pressure distribution on the wall of the toroidal chamber (b) S - *measurement area;* L - *load area.*

restoration of the magnetic pressure in the model was carried out using measurements of strains which were made in a narrow section of the bandage. The measurements of the strains were used to determine the axial stresses in the connector of the toroidal chamber subject to the force of a magnetic field. **Figure 17a** illustrates the axial stresses in the section of the bandage: «these define the error margin (curves 1–3) and are constructed on the basis of experimental data. The corresponding distribution of the magnetic field pressure on the inner contour of the bus is shown in **Figure 17b**, curves 1–3. These results are characterized by a satisfactory level of reconstruction and are consistent with the a priori ideas on the distribution of magnetic pressure.

It is proposed that zirconium bronze be used for the current-carrying elements that are subject to cyclic heating during the operation of the installation. In order to

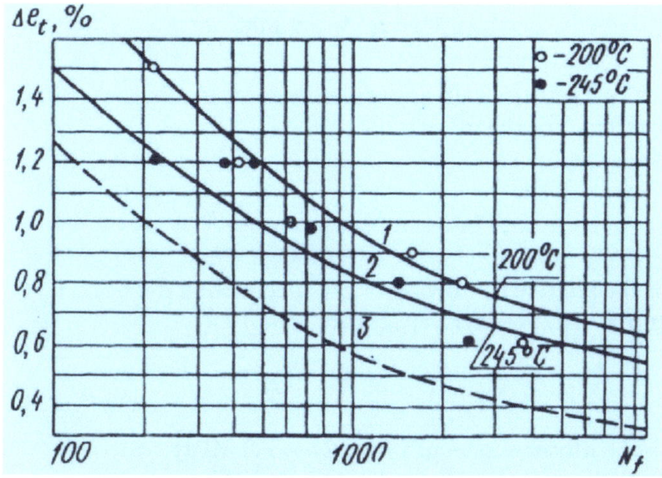

Figure 18.
Calculated low-cycle fatigue curves and experimental data for zirconium bronze at temperatures of 200 and 245°C.

determine its performance, its resistance to low-cycle deformation and fracture at elevated temperatures was studied.

The calculated low-cycle fatigue curves for zirconium bronze under rigid isothermal cyclic loading are shown in **Figure 18**, in which curve 3 represents the normative equation and characterizes the lower, conservative limit for low-cycle durability, with safety factors for strains $n_e = 2$ and for durability $n_N = 10$ [9, 10].

As part of the commissioning work on the creation of SFT, significant emphasis is given to experimental studies of the stress-strain state of the elements of the installation. One specific feature affecting the measurement of strain in the SFT unit is the presence of pulsed magnetic fields of up to 20 T, together with a change in the temperature of the current-carrying elements of up to 250°C and a high level of measured strains (up to 1.2%).

In connection with the impulsive growth of strains and the possibility of the transition of the elements of the wedge part of the bandage and the bus to a plastic state (see **Figures 15** and **16**), it seems possible to use the brittle strain-sensitive coatings method for the purpose of studying the stress fields. Estimates of mechanical stresses in the steel bandage of the model were carried out using the brittle strain-sensitive coatings method at magnetic fields reaching 14 T, with a yield point - of about 12 T.

The analysis and modeling of deformation processes of the elements of a thermonuclear installation in operating mode showed that fretting fatigue arises on the contact surface between the bandage and the bus as a result of the difference in displacements and the presence of contact interactions. This requires special study, since according to [7, 8] it has a significant impact on the installation's integral strength and service life parameters.

As noted above, the characteristics of the mechanical properties of bronze were determined by experiment, using specimens from supplied semi-circular forgings with an average radius of 270 mm and a cross-section of 110 × 140 mm. In order to carry out cyclic tests under conditions that simulate the operation of the busses in contact with the bandage at room temperature, a special loading device was developed that creates a transverse load on the specimen.

When quantifying parameters through the characteristics of the mechanical properties of the studied zirconium bronzes which are included in the Eqs. (1)–(3), this equation takes the form [2–4, 9].

$$e_a = 0{,}25e_c \cdot N^{-m_p} + 0{,}435\frac{S_c}{E}N^{-m_e}, \quad (8)$$

where $e_к$ is the fracture strain under a monotonic loading, $m_p = 0.65$, $m_e = 0.06$ are the characteristics of the material.

The value is equal to

$$e_c = \ln\frac{1}{1-\psi_c}, \quad (9)$$

where ψ_c is the relative narrowing of the specimen in the neck, $\psi_c = 0.7 \div 0.75$.

Tear fracture resistance at the specimen neck

$$S_c = \sigma_u\left(1 + 1{,}4\psi_c\right), \quad (10)$$

where σ_u is the ultimate strength (σ_u = 310–330 MPa),

E is the elasticity modulus ($E = (1.3 \div 1.5) \cdot 10^5$ MPa.

From the data obtained from service life assessments, taking into account the fretting effect, the amplitude of the fracture strains e_a in the Eq. (8) needs to be reduced by the reduction factor K_k. This factor reaches values of 2–2.25.

$$e_{ak} = e_a / K_к \quad (11)$$

which is comparable with the safety factor for durability $n_e = 2$, applied in the norms for nuclear reactor calculations [9, 10].

7. Conclusions

A combination of various experimental and theoretical studies to determine the load exposure and strength of tokamak installations is an essential foundation for the design of thermonuclear installations, for both demonstration and industrial purposes, and underlies the system of calculations and experiments required in order to actively monitor the stress-strain and limiting states of all load-bearing elements, taking into account the entire range of effects resulting from design, technological and operational factors. The resulting calculations and experimental data on temperature, stress, strain and displacement fields need to be included as initial components in assessments of the strength, service life and survivability of the load-bearing structures of new thermonuclear installations. In general, the research results summarized above demonstrate the correctness of the adopted design solutions, which - in the modes considered during the research - result in a relatively low level of local stresses in the load-bearing elements of the installations. This will contribute [1–3, 6, 23, 24, 26] to improvements in social and economic efficiency and overall safety during the transition from demonstration thermonuclear installations to industrial production, which is anticipated in the second half of the 21st century.

Acknowledgements

This work was financially supported by the Russian Science Foundation (grant no. 20-19-00769).

Author details

Nikolay A. Makhutov, Mikhail M. Gadenin, Sergey V. Maslov, Igor A. Razumovsky and Dmitry O. Reznikov*
Mechanical Engineering Research Institute, Moscow, Russia

*Address all correspondence to: mibsts@mail.ru

References

[1] Energy Strategy of the Russian Federation until 2035, Order of the Government of the Russian Federation No. 1523-r, dated 09.06.2020 (in Russian)

[2] Velikhov E.P., Smirnov V.P. State of research into and future prospects of thermonuclear energy. Herald of the Russian Academy of Sciences. 2006, Volume 76. No. 5. pp. 419-426 (in Russian).

[3] David Beltran. ITER Electrical Design Handbook. Earthing and Lightning Protection. ITER Technical Report ITR-20-007. The ITER Organization. July 16, 2020. 38 p.

[4] N.A. Makhutov, M.M. Gadenin, I.A. Razumovskiy, S.V. Maslov, D.O. Reznikov Probability Modeling Taking into Account Nonlinear Processes of a Deformation and Fracture for the Equipment of Nuclear Power Installations. /Combinatorics and Probability. Chapter 8. London. IntechOpen, 2019. pp. 191-220. DOI: http://dx.doi.org/10.5772/intechopen.88233

[5] N.A. Makhutov,K.V. Frolov,Yu.G. Dragunov and others. Risk Analysis and safety improvement in relation to water cooled power reactors. Moscow. Znaniye publ. 2009, 499 p. (in Russian).

[6] N.A. Makhutov Structural integrity, service life and technogenic safety. Book 2. Novosibirsk: Science publ. 2005. 610 p. (in Russian).

[7] N.A. Makhutov. Strength, service life, survivability and safety of machines. Moscow. LIBROKOM Publ. 2008. 576 p. (in Russian).

[8] Problems of strength, technogenic safety and structural materials science Ed. N.A. Makhutov, Yu.G. Matvienko, A.N. Romanov. – Moscow. LENAND publ, 2018. 720 p. (in Russian).

[9] Standards for calculation of durability of equipment and pipelines of nuclear power installations (PNAE G-7-002-86 - Rules and regulations in the nuclear power industry). Moscow. Energoatomizdat publ. 1989. 525 p. (in Russian).

[10] Boiler and Pressure Vessel Code. 2010 Edition. NY: ASME. 2011.

[11] ITER, International Thermonuclear Experimental Reactor. http://www.iter.org.

[12] N.A. Makhutov, M.M. Gadenin. Study of the service life of elements of a thermonuclear power installation, taking into account tribophysics parameters. Mechanics of machinery, mechanisms and materials. 2017. 3 (40). pp. 13-20 (in Russian).

[13] K. Miya, A. Kobayash, K. Koizumi, K. Hada, T. Shimakawa Construction of structural design guidelines for vacuum vessels and other components. Fusion Engineering and Design. 41. 1998.

[14] A. Raffray, G. Federici, V. Barabash, H. Pacher, H. Bartels, A. Cardella, R. Jakeman, K. Ioki, G. Janeschitz, R. Parker, R. Tivey, C. Wu. Beryllium application in ITER plasma facing components. Fusion Engineering and Design. 37 (2). 1997. pp. 261-286.

[15] B. Odegard, C. Cadden, N. Yang Beryllium-copper reactivity in an ITER joining environment. Fusion Engineering and Design. 41 (1-4) 1998. pp. 63-71

[16] F. Moons, R. Chaouadi, J. Puzzolante Fracture behavior of neutron irradiated beryllium. Fusion Engineering and Design. 41 (1-4), 1998. pp. 187-193

[17] G.P.Cherepanov. Methods of Fracture Mechanics: Solid Matter Physics. Springer. 1997. 335 p.

[18] Fracture mechanics and strength of materials. Reference book in 4 volumes. ed. V.V. Panasyuk. Kiev Naukova dumka publ. 1988-1990. (in Russian).

[19] V.D. Kuliev Singular boundary problems. Moscow FIZMATLIT publ. 2005. 719 p. (in Russian).

[20] E.M. Morozov, G.P. Nikishkov The finite element method in fracture mechanics. URSS Publishing house. 2008. 254 p. (in Russian).

[21] S.N. Atluri Computational Methods in the Mechanics of Fracture. North-Holland. 1986. 414 p.

[22] V.Z. Parton, E.M. Morozov Mechanics of Elastic-Plastic Fracture: Fundamentals of Fracture Mechanics. KomKniga publ. 2008. 400 p. (in Russian).

[23] ITER Design Description Document (DDD), G 16 DDD 2 96-11-27 W0.2.

[24] A. Cardella, A. Lodato, H. Pacher, R. Parker, K. Ioki, G. Janeschitz, D. Lousteau, A. Raffray, M. Yamada, C. Gusic. The ITER port limiter design. 43 (1), 1998, pp. 75-92.

[25] M. Williams. The stresses around a fault or crack in dissimilar media. Bull. Seismological Society of America. Bulletin of the Seismological Society of America. 1959. 49 (2). pp. 199-204.

[26] V.D. Kuliev, S.E. Bugaenko, I.A.Razumovsky Development of design criteria for ITER multilayer components. Brittle fracture of multilayer materials. Thermonuclear fusion. Collection of articles. Moscow. NIKIET publ. 1998. (in Russian).

Section 5

Nuclear Materials for Space Applications

Chapter 6

Nuclear Thermal Propulsion Reactor Materials

Douglas Burns and Stephen Johnson

Abstract

Nuclear thermal propulsion (NTP) systems have been studied in both the USA and the former Soviet Union since the 1950s for use in space science and exploration missions. NTP uses nuclear fission to heat hydrogen to very high temperatures in a short amount of time so that the hydrogen can provide thrust as it accelerates through an engine nozzle. Benefits of NTP systems compared to conventional chemical and solar electric powered propulsion systems include higher fuel efficiency, greater mission range, shorter transit times, and a greater ability to abort missions and return to Earth in the event of system failure. As a result of these benefits, the US National Aeronautics and Space Administration (NASA) is evaluating NTP for use in crewed missions to Mars, and plans for a possible mid-2020s flight demonstration of a NTP engine are under development. The extremely harsh conditions that NTP systems must operate in present a number of significant engine design and operational challenges. The objective of this chapter will be to describe the history of NTP material development, describe current NTP material fabrication and design practices, and discuss possible future advances in space propulsion material technologies.

Keywords: space, nuclear power, nuclear fuel, high-temperature materials

1. Introduction

The dream of one day expanding humanity's presence into the solar system will require advanced propulsion systems that provide high levels of thrust and efficient use of fuels. Thrust will be needed to leave Earth's gravitation field and to establish stable orbits when approaching other planets and returning home. Many of the missions that will one day be of interest to human explorers will require travel to locations that are far away from the sun, so dependence on solar power will not be an option, and prepositioning enough chemical propellant to allow freedom of movement and the ability to return to Earth will be too expensive.

A wide range of studies, including the National Aeronautics and Space Administration's (NASA) recent Design Reference Architecture (DRA) 5.0 Study [1], have shown that nuclear power can enable exploration of the solar system. Nuclear thermal propulsion (NTP) and nuclear electric propulsion (NEP) are technologies that can provide the necessary thrust and power densities to enter and leave gravity wells of planets, moons, and large asteroids, and they do not need external sources of power to generate propulsion. Heat produced through fission is all that is needed to add energy to a propellant and produce thrust.

Space nuclear reactors rely on nuclear fuels that include a range of fissionable compounds. Uranium oxide (UO_2), uranium nitride (UN), uranium carbide (UC and UC_2), and uranium oxycarbide (UCO) are ceramic materials that have been studied by various space reactor technology development activities. Each of these materials has advantages and disadvantages related to use in space reactors, but they are all capable of achieving the extremely high temperatures that will be needed to move humans and equipment from Earth to other parts of the solar system.

1.1 Fundamentals of rocket propulsion

The function of a rocket engine is to provide a force $\vec{F}$ over a time t to a body of mass m in order to change velocity $\vec{v}$ of the body by an amount $\overrightarrow{\Delta v}$. The rocket expends a mass Δm of fuel in order to complete a velocity change maneuver. The force on the engine is produced by heating a propellant and expelling it through an expansion nozzle at a velocity v_e with respect to the engine. The force produced is given by $F = \frac{dm}{dt} v_e$, where $\frac{dm}{dt}$ is the propellant mass flow rate.

The efficiency of an engine is determined by the force produced by a unit of mass flow rate, which is frequently defined in terms of "specific impulse." Specific impulse is given by $I_{sp} = \frac{v_e}{g}$, where g is the acceleration of gravity (note that I_{sp} has units given by *velocity* $\div$ *acceleration* $=$ *seconds*). During a maneuver, the initial mass of the engine m_o changes to a final value of m in order to produce a change in velocity $\overrightarrow{\Delta v}$, so that the mass ratio $\frac{m}{m_o}$ is a measure of maneuver efficiency. In free space, with no other forces acting on the engine, conservation of momentum leads to the "rocket equation" given by:

$$\frac{m}{m_0} = e^{-\frac{\Delta v}{v_e}} = e^{-\frac{\Delta v}{g I_{sp}}} \quad (1)$$

This equation illustrates how I_{sp} is tied to engine efficiency.

Another important aspect of rocket engine operations is that propellant exhaust velocity v_e is given by:

$$v_e{}^2 = \frac{k \frac{R}{M} T_c \left[1 - \left(\frac{p_e}{p_c}\right)^{\left(\frac{k-1}{k}\right)}\right]}{(k-1)} \quad (2)$$

where k is a constant given by the ratio of propellant liquid and vapor phase specific heats, R is the universal gas constant, M is the propellant molecular weight, T_c is the combustion chamber temperature, p_e is the nozzle exit pressure; and p_c is the combustion chamber pressure.

As a result, $v_e{}^2 \propto \frac{T_c}{M}$, and therefore $I_{sp} \propto \sqrt{\frac{T_c}{M}}$, so that engine efficiency increases in systems that produce high temperatures and use low molecular weight propellants. In chemical rockets, the highest available I_{sp} is produced by burning H_2 and O_2 to produce H_2O with a molecular weight of approximately 18 g/mol. Nuclear rockets, on the other hand, use H_2 as a propellant, so they produce specific impulses that are approximately $\sqrt{\frac{18}{2}} = 3$ times higher than the impulses produced by chemical rockets, for a given chamber temperature. **Figure 1** shows a comparison of theoretical specific impulses and mass ratios (i.e., ratio of take-off mass to final mass for Earth escape) for various propulsion systems [2].

Nuclear thermal propulsion systems can use a range of fluids for thrust and reactor cooling. Examples include hydrogen, ammonia, methane, octane, carbon dioxide, water, and nitrogen [3]. Specific impulse is lower for higher molecular weight fluids, but the heavier fluids require less storage capacity, and they could be mined, or synthesized, on interplanetary trips.

Nuclear engine design requires iterative consideration of reactor neutronic thermal hydraulic and structural characteristics combined with engine system-level performance analysis [4]. Effective design and analysis sequences involve establishing a preliminary core design that meets the fundamental neutronic performance requirements of start-up criticality and reactor control. Fuel element designs using fixed fuel compositions and uranium enrichments are developed early in the design process, and then the preliminary design is used to determine neutron and gamma energy deposition characteristics that feed an integrated thermal hydraulic/structural analysis of the core's internal components. Once acceptable neutronic and thermal/structural performance is achieved, overall engine performance is evaluated to determine how well the design satisfies mission requirements. The analysis sequence is then revised as necessary to optimize engine performance characteristics to support specific mission profiles.

Engine performance can be improved by various methods of controlling propellant flow through the reactor core and varying fuel compositions. For example, enrichment zoning within the fuel elements, with lower enrichments in high-power regions of the core, can be effective at flattening reactor power profiles and producing more uniform propellant exit temperatures. The cost of these design complications is often slightly reduced core reactivity that can have an impact on engine performance (i.e., specific impulse), but compensation for the reactivity loss is

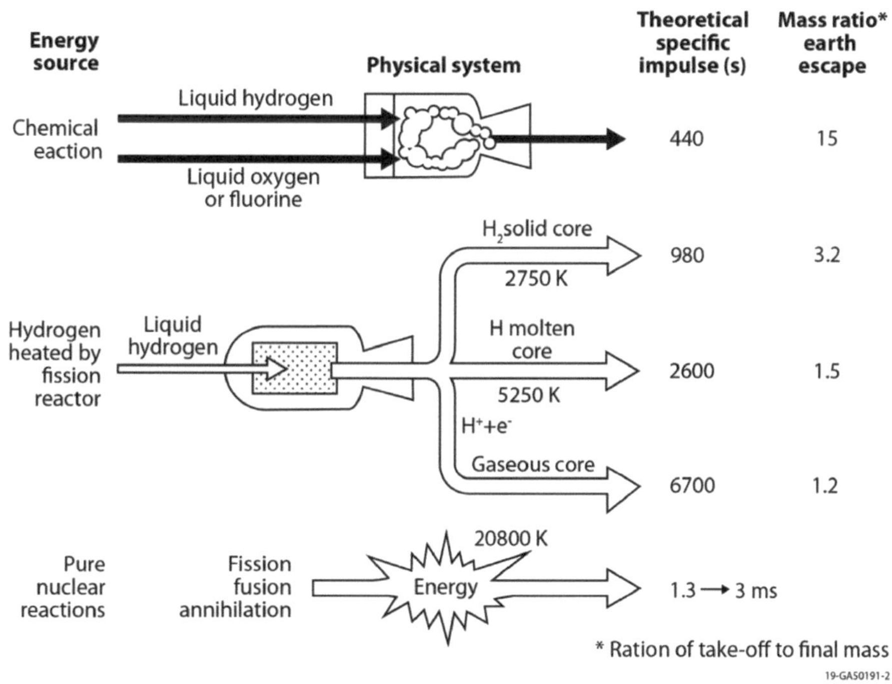

Figure 1.
Comparison of rocket propulsion system characteristics.

often possible through careful consideration of performance enhancements outside of the reactor fuel (e.g., propellant orificing, reductions in reactor mass, and the use of materials with low neutron absorption characteristics).

A wide variety of fast spectrum and thermal spectrum reactor designs have been developed for use in space propulsion systems. Fast spectrum reactors rely on high-energy (i.e., "fast") neutrons having average energies greater than 0.5 MeV to produce heat using materials that can fission after fast neutron absorption, while thermal spectrum reactors require the use of moderator materials to slow neutrons down to lower energies that are more readily absorbed. Fast reactors require fuel that is relatively rich in fissile material, while thermal reactors can operate with low-enriched uranium fuels.

Both fast and thermal spectrum reactors are typically designed with reflectors made from materials such as beryllium that prevent neutron leakage from the reactor core without producing a significant amount of neutron absorption. In space reactors, axial reflectors are often placed above and below the reactor core and radial reflectors are often placed around the core to reflect neutrons that would otherwise escape from the core back into the reactor's fuel. Control drums that are rotated to add enough reactivity to start up the reactor and make minor adjustments to its power profile are typically placed inside the radial reflector. A material with a high neutron absorption cross section (e.g., boron carbide, B_4C) is placed on one side of the control drums to remove neutrons while the reactor is shut down. The drums are rotated to move the neutron absorption material farther away from the core in order to start up the reactor.

Fuel depletion and fission product buildup during reactor operation are typically areas of concern for reactor design, but space reactor operating times are typically very short, so fuel burnup and fission product buildup are usually of little importance to NTP reactor designs.

2. Space reactor research and development programs

2.1 Rover/NERVA

Nuclear thermal propulsion systems were studied extensively during the 1950s and 1960s, but they were considered to be too heavy and expensive for deployment. At the time, chemical rockets and solar power were more economical for near-Earth operations that were the focus of the world's space agencies. However, recent interest in deep space exploration, and especially interest in sending astronauts to Mars, has reinvigorated NTP research for several reasons. First, the longer thrust duration than chemical rockets that can be produced with an NTP system could cut the travel time to Mars by 20–25%. The reduced travel time is important because it would reduce the amount of dose that astronauts would receive from cosmic radiation during the voyage. Second, the higher thrust would extend the available launch window for missions to Mars. Conventional chemical rocket engines can only be used to reach Mars during a 30-day window that opens every 26 months due to the relative positions of the planets, while nuclear propulsion systems can provide enough thrust to leave Earth's orbit and reach Mars during more points during the orbital profiles of the two planets. Finally, NTP systems could extend the amount of time during which a mission to Mars could be aborted and still allow astronauts to safely return to Earth. Chemical rocket fuel needed for the return trip from Mars to Earth would likely need to be sent to Mars in advance of a crewed mission, so astronauts would have to reach Mars in order to return to Earth after their initial supply of fuel is consumed, if chemical rockets were used for the mission. An NTP system, on the other hand, would be able to turn around and return to Earth before reaching Mars, if the mission had to be aborted.

Several successful research and development programs focused on space reactor fission power technologies have been established over the past 60 years. The earliest, and most extensive of these efforts, were the Rover and Nuclear Engine for Rocket Vehicle Application (NERVA) programs that were sponsored by the US Atomic Energy Commission (AEC) between 1958 and 1971 [5]. A total of 13 research reactors and 6 nuclear engines were built and tested under the Rover/NERVA programs at the AEC's Nevada Test Site (NTS) Nuclear Rocket Development Station (NRDS) and other facilities located across the country [6]. The Rover reactor development and testing efforts were led by the Los Alamos Scientific Laboratory (LASL), and the NERVA reactors were designed and built by Westinghouse Electric Corporation (Astronuclear) and Aerojet-General Corporation following a 1961 design competition. The Kiwi (1955–1964), Phoebus (1964–1969), and Peewee (1969–1972) series of reactors were developed and tested under Rover to demonstrate the basics of nuclear rocket technology and to study characteristics of high-temperature nuclear fuels and long-life fuel elements. The NERVA NRX and XE engines were also built between 1964 and 1969 and tested at NRDS to study the complexities of nuclear engine start-up, full-power operation, and shutdown. A list of the best performance parameters achieved during the Rover/NERVA programs is presented in **Table 1** [7].

Early work in the Rover/NERVA program was performed with uranium carbide (UC or UC_2) and uranium dioxide (UO_2) fuel particles embedded into graphite. The effects of fission product interaction with the graphite matrix quickly led to the use of pyrolytic graphite-coated UO_2 particles in the Rover/NERVA fuel designs. The pyrolytic carbon contained fission products produced during reactor operations before the fission products could cause dislocations in the graphite structural materials. The pyrolytic carbon layer also protected the UO_2 particles from oxidation during fabrication and handling of the fuel.

A major drawback associated with using graphite in space reactors is that graphite converts to methane (CH_4) under hydrogen exposure, and the conversion causes the graphite to corrode. During the Rover/NERVA programs, the graphite matrix used in the reactor fuel was protected from corrosion through application of a high-temperature niobium carbide (NbC) coating. However, the use of the NbC coating leads to issues associated with "mid-band corrosion," which was higher corrosion rates in the center third of the reactor fuel elements where power density

Parameter	Reactor (test date)	Peak performance achieved
Power	PHOEBUS 2A (July 1968)	4100 MWt
Peak fuel temperature	PEEWEE (November 1968)	2750 K
Specific impulse	PEEWEE (November 1968)	848 s
Maximum restarts	XE' (June 1969)	28
Accumulated time at full power	NF-1 (June–July 1972)	109 min
Continuous operation	NRX-A6 (December 1967)	62 min

Table 1.
Maximum performance results achieved during the Rover/NERVA programs.

was high. The corrosion was found to be caused by cracking of the NbC coatings used in the fuel's hydrogen coolant channels due to a mismatch between the thermal expansion coefficient of the NbC coating and the thermal expansion coefficient of the fuel matrix. The NbC coatings were eventually replaced with zirconium carbide (ZrC) coatings because of zirconium carbide's superior resistance to fission product diffusion at high temperatures.

The search for solutions to the corrosion problem also led to the development of {(U, Zr) C, graphite} composite fuel. The coefficient of thermal expansion for the composite fuel was 6–6.5 μm/mK (versus 3 μm/mK for fuel made from UO_2 dispersed in graphite), which matched the NbC coefficient of 7.1 μm/mK fairly well [7].

Carbide material systems have several favorable features applicable to NTP reactors including:

- Relatively small neutron absorption cross sections
- High melting points
- Thermal stability
- Low volatility
- High fuel densities
- High moderation ratios
- Low material densities (<10 g/cm^3)

However, the complex fuel designs used in the early Rover/NERVA tests were challenged by the mechanical loads, thermal stresses, and high radiation fields found in NTP reactors. The intense operational conditions contributed to the formation of stress fractures in the fuel coatings and surfaces and the cracking encouraged increased hydrogen penetration into the fuel that produced fuel degradation.

The Small Nuclear Rocket Engine (SNRE) was the last engine design studied by Los Alamos National Laboratory (LANL) under the Rover/NERVA program. The SNRE was a nominal 16,000 lbf thrust engine that was originally intended for short run-time, unmanned missions, and the SNRE stage design was constrained to fit within the payload volume of the planned space shuttle. The reactor's core design used hexagonal fuel elements and hexagonal structural support elements (i.e., tie tubes), and the number of elements could be varied to support different thrust requirements. Higher thrust designs for SNRE meet or exceed performance characteristics identified in the DRA 5.0 study, so SNRE would be suitable for use in human missions to Mars.

2.2 GE-710 high-temperature gas reactor research and development program

The GE-710 [8] high-temperature gas reactor (HTGR) and the Argonne National Laboratory (ANL) nuclear rocket engine programs [9] focused on development of ceramic-metal (cermet) fuels consisting of uranium ceramic material (e.g., uranium dioxide [UO_2] or uranium nitride [UN]) embedded in a refractory metal matrix (e.g., tungsten). To ensure good bonding between the kernels and the matrix, the kernels were coated with a thin layer of the matrix metal (i.e., tungsten [W] or molybdenum [Mo]). In addition, the coolant flow channels of the cermet fuel were coated with either tungsten or niobium.

The GE-710 program ran from 1962 to 1968 with the objective of performing reactor tests of a closed-loop system (i.e., an engine system that recycled engine propellant) that used neon as a coolant, and an open-loop system (i.e., an engine system that expelled the reactor coolant to produce thrust) that used hydrogen as the reactor coolant. Final program goals focused on longer-term operation (approximately 10,000 h) at fuel temperatures in the 2000–2250 K range. Major achievements during the GE 710 program included down selection to either W-UO_2 or Mo-UO_2 cermet fuels, significant development of fabrication and brazing techniques for cermet fuel elements, development of sintering methods for fabrication of high-density fueled cermets, and initiation of in-pile testing. Molybdenum was also investigated as a substitute for the tungsten matrix, but the lower strength of Mo caused increased fuel swelling at high burnups due to fission gas buildup. The loss of Mo due to vaporization at high temperatures during electron beam welding and during thermal cycling was also undesirable.

The W-UO_2 cermets tested under the GE-710 HTGR program were cold pressed and sintered into segments of approximately 12.7 mm lengths. Tungsten powder composed of 1–2 μm diameter particles with a uniform spherical shape was used in the sintering process. These particles produced increased sintered densities (e.g., 95% of theoretical density) compared to coarser particles with predominantly angular or planar shapes. Microspheres are a desirable particle shape because they have good heat transfer characteristics, they have consistent grain sizes, they are non-abrading (and therefore dust free), they are free-flowing, and they can be engineered to be either soft or hard. Conversely, powders with varying grain sizes are undesirable because they tend to agglomerate, they can be abrasive, and they have low reproducibility.

After sintering, the GE-710 fuel segments were machined into a hexagonal shape. Coolant channels with 0.914 mm (0.036 in) diameters were drilled into the segments, and then 0.203 mm (0.008 in) wall thickness coolant tubes were sealed into place on one end of the segments by tungsten inert gas (TIG) welding. A header was brazed to the other end prior to complete element assembly, and a tantalum (Ta) spacer plate was used adjacent to the header to protect the fuel from the braze material. The segments were placed into a 0.381 mm (0.015 in) wall thickness cladding, and the fuel was bonded to the cladding using a high-temperature, high-pressure autoclaving process. Autoclaving was typically carried out at a pressure of 10.3 megapascals (MPa) (1494 pounds per square inch [psi]) and 1922 K for 1 h, although an alternate hot-gas pressure process was also used at 68.9 MPa (9993 psi) and 2022 K for 2–3 h.

Dissociation of UO_2 into free uranium and hyperstoichiometric UO_2 or oxygen during the sintering process had a detrimental effect on fuel fabrication. Fuel performance issues arose from UO_2 dissociation because an increase in excess oxygen within the fuel led to an increase in fuel swelling, and the excess oxygen could react with the W matrix to form WO_2 stringers in the matrix grain boundaries. Free uranium was also detrimental to fuel dimensional stability and caused negative reactions with the cladding materials that were used. There was little to no mobility of free uranium below the fuel particle melting temperature range (1422–1644 K), but uranium formed a two-phase mixture that produced fuel swelling above the melting temperature range.

Challenges associated with dissociation of UO_2 were first addressed by the addition of thorium oxide (ThO_2) as a stabilizing compound, but testing showed that ThO_2 only delayed free uranium migration. A more suitable solution to UO_2 dissociation was found to be the addition of substoichiometric UO_2 combined with the ThO_2 stabilizer since substoichiometric UO_2 retained a single phase as temperatures increased. Dissociation and free uranium migration became an issue only during thermal cycling, and oxygen to uranium ratios of 1.984–1.988 were found to have the best performance during thermal cycling tests [10].

Differences in the coefficient of thermal expansion between the fuel, matrix, and cladding also presented themselves during the development program. During bonding of the cladding to the fuel, the cladding material expanded two times more than the matrix material, resulting in compression at the interface when the fuel was cooled. Alloying with 3 wt% rhenium (Re) in the tungsten matrix increased the low-temperature ductility of the matrix.

The most promising clad materials used in the GE-710 program were elemental tantalum, tantalum alloys (T-111 and tantalum-10 weight percent tungsten [Ta-10W]), and a tungsten-30 weight percent rhenium-30 weight percent molybdenum (W-30Re-30Mo) alloy. Tantalum was selected as the initial cladding material because the material was readily available, and it had sufficient compatibility with the W-UO_2 cermet fuel. However, tantalum clad performance was limited by free uranium that formed reaction voids in the cladding. The voids formed because repeated cycling of the fuel allowed uranium metal to precipitate out of single-phase UO_{2-x}. The uranium migrated through the W matrix grain boundaries and into the Ta cladding, and leak paths developed as the uranium metal re-oxidized to form UO_2. The T-111 cladding material was attractive because the alloy maintains a fine-grained structure that limits uranium movement until grain growth occurs above 1922 K. However, the alloy has a high oxygen permeability that results in reaction void formation. The W-30Re-30Mo alloy used in the later stages of the GE-710 program was found to have low oxygen permeability, low sensitivity to gas impurity absorption, high strength, high melting point, and good ductility. Unfortunately, high bond stresses caused by thermal expansion mismatch between the W-30Re-30Mo clad and the fuel matrix occurred during thermal cycling. An anneal heat treatment was used to overcome the bond stresses, but the treatment caused re-precipitation of the sigma phase at grain boundaries which led to clad weakness. Volatilization of Mo at high temperatures also increased sigma phase formation and reduced clad strength.

Irradiation tests performed under the GE-710 program included tests of UO_2 and ThO_2-stabilized UO_2 fuel samples in the Idaho National Laboratory's (INL) Engineering Test Reactor (ETR). Matrix materials used in the samples included W, W-Re, and Mo with Ta-10W, W-30Re-30Mo, and niobium (Nb) cladding. Approximately half of the samples evaluated in the ETR testing campaign developed fission gas leakage. Further testing was performed in the Low-Intensity Test Reactor (LITR) at the Oak Ridge National Laboratory (ORNL) using W and W-Re matrix material with W-30Re-30Mo and W-25Re-3Mo cladding. The results of the test were similar to the ETR results. However, a third series of tests in the Oak Ridge Research (ORR) reactor with basically the same matrix-cladding combinations showed significant improvements that were achieved by reducing the density of UO_2 in the fuel to provide void space for fission product gas accumulation.

Fuel failure modes observed during the GE-710 testing included [7]:

- Loss of oxygen from UO_2 at high temperatures followed by the formation of substoichiometric UO_2, free uranium, and uranium penetration of the cladding wall during thermal cycling.

- Volume expansion, and eventual cracking of the W-UO_2 fuel matrix, during very high-temperature operation after significant thermal cycling.
- Void formation between the cermet fuel and the fuel cladding during fabrication and early operation.
- Fission product damage/release after 4000–7500 h of operation at 1870–2270 K in fuel specimens sintered to 95% or greater theoretical densities.
- Preferential vaporization of Mo and other lower melt point materials out of the clad at temperatures above 2470 K. Molybdenum was found to be a poor candidate for alloying because Mo vapor pressure becomes significant at temperatures above 2470 K.

Physical mechanisms determined to have caused the failure mechanisms included:

- Transparency of Ta and Ta alloy materials to oxygen at intermediate and high temperatures.
- Volume expansion and cracking caused by incompatibility of coefficients of thermal expansion between tungsten and UO_2 in the fuel matrix. The incompatibility caused fuel particles to pull apart from the tungsten matrix at high temperatures and after multiple thermal cycles.
- Void formations caused by difficulties with achieving good seals between metal alloy cladding, internal metal alloy coolant tubes, and the cermet fuel material.
- Insufficient permeation of alloy-clad material into the cermet during autoclaving, leaving weaknesses that developed into voids.
- Fission product damage to the cermet, and eventually the cladding material, caused by accumulated buildup of pressure, lattice stresses, and dislocation weaknesses under irradiation.

Sintering to lower theoretical densities of 84–90% created a significant improvement in sample performance. An increase in burnup capability (i.e., fissions/cm^3) by almost a factor of 10 was achieved by simply giving the fission products additional room for expansion without exerting stresses in excess of the tungsten matrix capability at elevated temperatures.

2.3 Argonne national laboratory nuclear rocket engine research and development program

The ANL nuclear rocket program focused on developing two reference reactor designs; the ANL200 and ANL2000 reactors were 200 MWt and 2000 MWt fast spectrum thermal propulsion systems that were designed to produce 44.5 kN (10,000 lbf) and 445 kN (100,000 lbf) of thrust.

Most of the ANL program's work was focused on design and testing of the ANL2000 system. The reactor consisted of an array of 163 hexagonal fuel elements that were assembled into an approximately cylindrical core with a diameter of 66 cm (26 in). The fuel elements were made from a 93% enriched tungsten-urania cermet fuel that was clad with 0.76 mm (0.03 in) of a tungsten-rhenium (W-25Re) alloy. The elements had a total length of 130 cm (51.56 in) and a fueled length of 87 cm (34.25 in). The core was supported from an Inconel Inco-718 grid plate that was bolted to the reactor vessel at the cold end of the core, and a cylindrical beryllium-oxide axial reflector containing 12 control drums was mounted at the inlet end of the core. A preheater consisting of stainless steel-UO_2 fuel elements at the inlet side of the reactor was also included in the reactor design.

The ANL2000 development program's performance goals were to reach a fuel temperature of at least 2770 K in order to produce an Isp of 821–832 s, achieve 10 hours of operation with at least 25 thermal cycles, and limit fuel loss to less than 1%. All of the program's goals were achieved before the program was terminated; however, neither of the ANL program's reference reactor designs were built or tested before the program was cancelled.

The primary fuel evaluated under the ANL program was UO_2 embedded in a tungsten matrix. The fuel choice was similar to the GE-710 program, but gadolinia was used to stabilize the ANL fuel, in contrast to the ThO_2 that was used in the GE-710 program. Three fuel fabrication methods were investigated under the program: cold pressing and sintering of W-UO_2 wafers, isostatic sintering of long fuel elements, and hot pneumatic compaction.

The cold pressing and sintering technique led to fabrication of approximately 6.3 mm (0.02 in) thick W-UO_2 wafers that were stacked to form a fuel column. Fuel grading was used in the stacks to optimize physics and thermodynamics of the core. The fuel fabrication method required a high strength cladding since the cladding provided structural support. The isostatic sintering method allowed for single-step fabrication of fuel elements that were approximately 45.7 cm (18 in) long. The process minimized concerns over coolant channel alignment tolerances because individual fuel wafers did not have to be stacked to form an element. Finally, the hot pneumatic pressing method was used to demonstrate the fabrication of fuel formed from UO_2 fuel kernels that were CVD coated with tungsten. A fuel loading of 60 vol% of 93% enriched UO_2 inside a W or W-Re matrix was used for all of the program's fuel samples.

Similar to the GE-710 program, stabilizers were added to the ANL program fuels to inhibit UO_2 dissociation, but the stabilizers investigated under the ANL program included gadolinium ($GdO_{1.5}$), dysprosium ($DyO_{1.5}$), yttrium ($YO_{1.5}$), and MoO_3. Ten mole percent of stabilizer was added to the UO_2 for all investigations.

The fuel fabrication process that gave the best results was a powder metallurgical process that produced near net shape fuels with cold isostatic pressures, followed by sintering at approximately 1500 K and chemical vapor deposition (CVD) of cladding on the coolant channels, even though deposition of uniform CVD coatings was difficult in the 1960s. The gadolinia stabilized fuel showed excellent retentivity at 2770 K for up to 45 hours and 180 cycles in non-nuclear tests performed in two hydrogen loops. Other tests showed that flowing hydrogen at temperatures exceeding 2700 K had essentially no impact on fuel loss rates.

Induction brazing was investigated by the ANL program as a means for joining fuel sections. A Zr-Mo braze with a melt temperature above 1973 K was the most successful; however gas generated during brazing made it almost impossible to fabricate a leak-free joint. The problem was overcome by immersing fuel sections in liquid nitrogen with the section to be brazed left above the liquid pool. Brazing was carried out in five sections to avoid allowing the fuel section to reach a temperature where volatilization of impurities could occur.

High-temperature refractory brazing techniques were also developed under the ANL program. Solid-state diffusion bonding of W-25Re alloys using nickel as an interleaf material that forms an Ni-W-Re ternary has been demonstrated at temperatures as low as 1173 K (although a temperature of 1773 K is required to produce sufficiently strong bonding). Brazing of refractory metals is generally undesirable due to recrystallization of microstructures produced in the joint, but solid-state diffusion bonding avoids recrystallization through the use of low temperatures. Nickel may be an undesirable interleaf material for high-temperature NTP materials, but other interleaf materials may be identified with further investigation [11].

Nuclear tests on the ANL cermet samples were performed in the Transient Reactor Test (TREAT) facility at INL. Eight cermet specimens, each with seven coolant channels and vapor-deposited tungsten cladding, were tested in the TREAT experiments. The test durations were typically 200–430 ms, although two samples were subjected to flat-top transients lasting 2–3 s. One of the tests failed, as fuel material was ejected from the sample, and the failure was attributed to fabrication issues, particularly tungsten coating thickness irregularities. The last two samples evaluated in the campaign were subjected to multiple transients at heating rates up to 16,000°C/s, a maximum temperature of 2870 K, and a power density of 30 MW/l. These samples showed no evidence of damage [7].

2.4 Space Nuclear Thermal Propulsion research and development program

The goal of fuel development under the Space Nuclear Thermal Propulsion (SNTP) Program was to develop a coated nuclear fuel particle with a diameter of approximately 500 μm that would support a mixed mean hydrogen exhaust temperature of 3000 K when incorporated into a particle bed reactor (PBR) [12]. The requirement gave a maximum fuel temperature target of 3100–3500 K based on a power density of 40 MW/l, whereas the maximum fuel temperature demonstrated during the Rover/NERVA program was in the range of 2400–2600 K.

The particle bed reactor concept developed by Dr. James Powell and his team at Brookhaven National Laboratory (BNL) caught the attention of SDI program managers as a possible power source for a rapid intercept vehicle that could destroy ballistic missiles, because it had the potential to overcome limitations associated with high-power production. Interest in the PBR technology led to the creation of the Timberwind program in 1987 and creation of the SNTP program in 1991, after Timberwind was declassified and transferred to the US Air Force.

The PBR fuel element designed for the SNTP program consisted of a large number of UC_2 fuel particles packed between two porous cylinders called frits. The fuel elements were housed inside cylindrical moderator blocks made of beryllium or lithium hydride that slowed the reactor's neutrons down to thermal energies that could sustain a fission chain reaction. Hydrogen served as both a coolant and propellant for the SNTP engine as it moved through the cold frit located on the outside of the fuel elements, flowed through the element particle beds to remove heat produced by the fission reaction, and then exited the fuel through the inner hot frit. The hydrogen then flowed axially down an annular channel located at the center of each of the core's fuel elements and exited the core before expanding through the engine nozzle to produce thrust.

The PBR concept promised significant reductions in system mass over solid core reactors due to the 20-fold increase in heat transfer surface area of the particle fuel elements compared to the prismatic fuel used in the Rover/NERVA program. PBRs also had a lower core pressure drop due to the shorter flow paths through the pebble beds. The small size of the particles helped to prevent cracking, because thermal gradients across the particles are relatively low, but the coatings used on the particles were found to be prone to high-temperature vaporization that was made worse by the high surface area to volume ratio of the particles.

The SNTP program began working on the development of coated fuel particles based on the HTGR Program fuel design. These particles were known as the program's baseline fuel. The baseline fuel development included the production of uranium-bearing fuel kernels using the internal gelation process. The fuel kernels were covered by pyrolytic carbon using chemical vapor deposition in a fluidized bed.

Babcock and Wilcox Inc. (B&W) developed the ability to produce ZrC outer coatings on microparticles with the assistance of LANL and General Atomics. B&W produced fuel particles consisting of UC_{2-x} kernels coated with two layers of pyrocarbon and an outer layer of ZrC that supported the Particle bed reactor Integral Performance Element (PIPE) experiments that were performed in 1988 and 1989. The first pyrocarbon layer in the fuel particles was a porous layer that accommodated the mismatch in thermal coefficient of expansion between the fuel kernel and the outer ZrC layer. The second layer was dense pyrocarbon that protected the fuel kernel from attack by the halides used in the CVD process. The outer ZrC layer was used to delay corrosion of the fuel kernel after it was exposed to hydrogen propellant.

More than 200,000 particles were tested in Sandia National Laboratory (SNL) Annular Core Research Reactor (ACRR) in four particle nuclear tests (PNT) [12]. Fuel temperatures achieved during the tests ranged from 1800 to 3000 K, and testing times ranged from 100 to 600 s. Baseline UC_{2-x} fuel kernel performance is limited by its melting temperature of 2700–2800 K, but the PNT tests showed that the melting temperature of fabricated UC_{2-x} kernels was actually closer to 2500 K. Molten UC_{2-x} dissolved the particle carbon layers and attacked the ZrC outer layer during the tests, and a complete particle failure occurred about 5 min after kernel melting. It is possible that increasing the graphite layer thickness would delay the time to failure, but the increased particle size might weaken any fuel matrix that was used to contain the particles, so testing of increased graphite layer thicknesses was not performed by the SNTP program.

The program pursued a dual fuel development path once it became clear that coated UC_{2-x} kernels would not meet the program's temperature requirements. Under the dual-path effort, BNL investigated the development of an infiltrated kernel (IK) fuel, and B&W investigated mixed-carbide fuel particles. BNL postulated that IK fuel could be formed when molten UC_{2-x} distributes uniformly through a porous graphite matrix. The laboratory's scientists reasoned that the molten uranium ceramic could be held within the graphite's pores and protected from hydrogen corrosion by an appropriate high-temperature outer layer, since UC_{2-x} is thermodynamically stable with respect to graphite and does not react with it even after melting. BNL demonstrated in 1992 at laboratory scale that molten UC_2 could be infiltrated into porous graphite coupons to the desired uranium density and that spherical IK particles could be fabricated. The demonstration also showed that pyrolytic layers used in the baseline fuel design are unnecessary in the BNL IK fuel, so IK fuel has a higher uranium density and smaller particle size than the baseline SNTP fuel.

The B&W mixed-carbide fuel design developed under SNTP was based on investigations that were performed at the end of the Rover/NERVA program. The fuel was formed as a mixture of refractory carbides such as ZrC, NbC, TaC, HfC, and UC. Uranium carbide has a theoretical melting temperature of 2798 K, but the refractory metal carbides have melting temperatures ranging from approximately 3700 K for ZrC to greater than 4200 K for TaC and HfC. Tantalum and Hf have relatively high neutron absorption cross sections, so only ternary mixtures of U-Zr-C and U-Nb-C were considered by the B&W fuel development program.

The diagram shown in **Figure 2** is an example of phase relationships for a mixed carbon fuel [13]. As illustrated in the figure, the melting temperature of mixed-carbide fuels decreases with increasing uranium content. The necessary uranium content for SNTP fuel was determined by fuel criticality conditions, and the B&W research

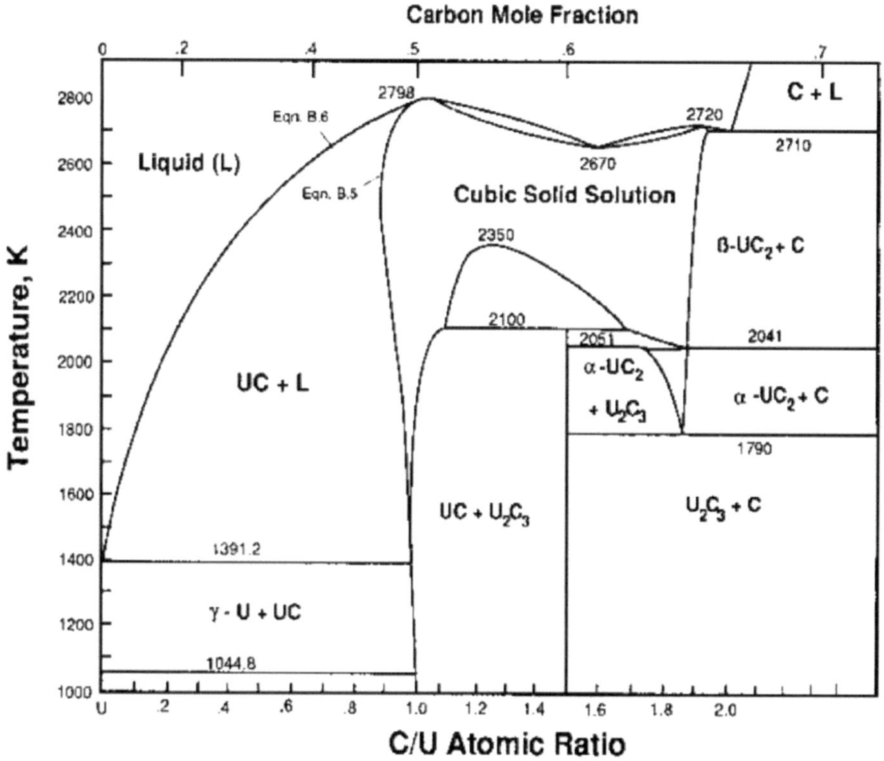

Figure 2.
Uranium-carbon phase diagram [13].

identified a minimum required uranium mole fraction of 0.15, which equated to a melting temperature of approximately 3200 K. By the end of 1992, B&W measured the melting temperature of U-Zr-C as a function of uranium content; measured the plasticity of ZrC, NbC, and U-Zr-C at 3200 K; and produced a small amount of NbC-coated U-Zr-C kernels using an internal gelation manufacturing process and CVD coating.

Overall accomplishments of the SNTP program included:

- Acquisition of technology and equipment that allowed production of nuclear fuel microparticles using the internal gelation process and coating of the particles with pyrolytic carbon and refractory metal layers using CVD
- Production of baseline fuel particles that supported radiation and non-radiation testing
- Development of a laboratory process for infiltrating porous graphite with uranium to produce infiltrated kernel particles
- Modification of the internal gelation process that allowed for production of U-Zr-C particles

2.5 Russian space nuclear engine research and development

The Union of Soviet Socialist Republics (USSR) performed a significant amount of research and development on nuclear thermal propulsion fuels from the 1960s to the late 1980s. Reported work included:

- Fabrication of a large number of samples
- Non-nuclear hot hydrogen flow tests
- Individual fuel element tests under rapid transient conditions in the Impulse Graphite Reactor (IGR)
- Prototypic NTR operating condition testing in the IVG-1 and RA reactor

The USSR followed the NERVA program quite closely and chose to follow the mixed-carbide fuel path early in its fuel development program. (U, Zr) C fuel was used for the low-temperature portion of the USSR reactor design (i.e., propellant exit gas temperature ≤ 2500 K), and (U, Zr, Nb) C was used for the high-temperature portion of the reactor core (i.e., propellant exit gas temperatures up to 3100 K). Some work using Ta and Hf in place of Nb was also reported. There were claims that Ta and Hf could produce 200 K higher fuel temperatures, but there was concern over the higher neutron capture cross sections of these elements compared to Nb. Finally, carbon nitride fuels were developed under the USSR program, primarily for use with ammonia propellants.

The USSR research program fabricated carbide fuels in a wide range of shapes, but the twisted ribbon geometry was the preferred fuel design. This geometry included long rods of fuel with many different cross-sectional shapes. The rods were twisted along their long axis and bundled together using wire wraps, or insertion into long canisters, to form fuel elements [14]. During the operation of the reactor, the propellant was directed through the bundles to transfer heat from the fuel. The twisted ribbon geometry provided a large surface area for heat transfer, and it could be fabricated in large volumes, although researchers from outside the USSR program were not allowed to observe the fabrication process.

Tests were performed on the USSR nuclear thermal rocket fuel design over a period of 19 years on approximately 1550 fuel assemblies. The testing program included seven full-core tests and approximately 160 transient tests that were performed at the IGR between 1962 and 1978 [7]. The highest reported hydrogen exit gas temperatures from testing performed during that period ranged from 2800 to 3300 K. Reported power densities were as high as 20 MW/l, and uranium loss estimates were as low as 0.5–1.0% based on reactivity loss measurements [15]. Very little postirradiation examination data on the fuel samples has been reported.

3. Carbon-based fuels and materials

The GE-710 and ANL programs were established as backups to the Rover/NERVA program. The choice of evaluating refractory metal-based fuels as a secondary fuel type to the Rover/NERVA graphite fuel research resulted from the greater experience base associated with graphite fuel, graphite's low thermal neutron absorption cross section, and the greater fabricability associated with graphite fuels.

Graphite was first used in nuclear reactors as a moderator, and large bars of polycrystalline graphite were used in many early reactors. A halogen purification process was developed to produce the high-purity graphite needed for natural uranium-plutonium production piles. More recently, graphite has been used as a fuel particle coating and as a matrix for fuel particles in high-temperature reactors [16].

The term "graphite" refers to a wide range of materials made from carbon that have a variety of properties. For example, graphite can be used as both a thermal conductor and a thermal insulator, it can be made in very dense and very light forms, and it can be highly anisotropic or isotropic. Graphite also has a wide variety of uses in nuclear reactor applications. It can serve as a high-purity neutron moderator, and it can be used in control rods and shielding with the addition of boron.

The variety of properties associated with nuclear grade graphite means that it can be difficult to obtain graphite that has specific properties within narrow limits that are consistent from batch to batch. New sources of graphite often have unknown property variations.

Graphite production processes are often proprietary, but the general method of manufacturing crystalline graphite includes [17]:

- Raw petroleum coke is calcined at 1300°C, milled, sized, and mixed at about 165°C with a coal-tar pitch.

- The mixture is cooled to 110°C and extruded.

- The extrusion is cooled to room temperature to form a "green body" and placed in a baking furnace supported by a permeable pack of sand and carbon.

- A large volume of gas evolves from pyrolysis of the pitch during baking to 800°C and the carbon body shrinks about 5 vol%.

- The material is then graphitized in an electric furnace at 2500–3000°C. Some further gas is vaporized during graphitization, but the principle physical change involves transformation from amorphous carbon into crystalline graphite.

There are many variations that can affect final material properties. For example, the baked carbon can be impregnated with pitch to increase density and strength, carbon black can be added to improve density and strength, and the graphitized body can be heated in a halogen-containing environment to remove trace impurities.

Pyrolitic carbon is made from decomposition of hydrocarbon gasses. For free-standing bodies, the carbon is usually deposited on a graphite substrate at temperatures from 1400°C to 2400°C. Material orientation, density, and other properties can be varied by changes in gas pressure, temperature, and other conditions. Subsequent heat treatment at higher temperatures can improve crystallinity, and small samples heat treated to 3000–3600°C (3273–3873 K) have shown electrical properties that are close to the properties of single crystals. Larger samples with near-single-crystal properties can be made by heating pyrolytic graphite above 2500°C (2773 K) under a compressive stress. Fuel particles are typically coated with pyrolytic carbon in a fluidized bed with the carbon coatings being applied to thicknesses of up to about 100 μm.

Carbon is a relatively light atom, so graphite is an efficient moderator. Slowing down power is the logarithmic energy change of a neutron when it collides with a moderator, and a nuclear graphite with a density of 1.65 g/cm^3 has a slowing down power of 0.063 cm^{-1}. Light water has the highest slowing down power of 1.5 cm^{-1}, and only several other materials such as beryllium (Be), beryllium oxide (BeO), and deuterium oxide used in heavy water reactors (D_2O) have higher slowing down power than graphite. Graphite also absorbs fewer thermal neutrons than any other material except D_2O.

Graphite is relatively weak at low temperatures, with a compressive strength of only a few thousand psi. However, its high-temperature strength is very good compared to other materials. Graphite's strength increases with temperature and reaches a maximum at about 2500°C (2773 K). A typical polycrystalline nuclear graphite with a tensile strength of 2000 psi at room temperature has a strength of about 4000 psi at 2500°C (2773 K). Graphite's high-temperature strength, good nuclear properties, and low cost are the primary reasons for its extensive use in high-temperature gas-cooled and nuclear propulsion reactors.

Carbide fuels such as UC and UC_2 have advantages over more widely studied oxide fuels. The most important advantage is their higher thermal conductivity, which approaches the value found in metallic uranium. Higher thermal conductivity lowers peak centerline fuel temperatures, which in turn allows for higher linear heat generation and larger diameter fuel rods. Carbide fuels also have higher uranium densities than UO_2, which allows for design of more compact reactors [15].

Mixed carbides such as uranium-zirconium carbide solid solution ([U, Zr] C) fuels have higher melting temperatures than UC. Research into mixed-carbide fuel fabrication has taken place in the USA and former Soviet Union to support space nuclear power applications. Three major carbide fuel designs were investigated under Rover/NERVA:

- UC_2 particles with pyrolytic carbide coatings and dispersion in graphite
- Composites of (U, Zr) C and graphite with the carbide forming a continuous web structure within the fuel
- Solid solution (U, Zr) C

All of the fuel designs, except the solid solution design, used protective ZrC coatings. Only 28 solid solution fuel elements were tested under Rover/NERVA, so effectiveness of the fuel design was not fully evaluated.

A solid solution is formed when two metals are completely soluble in their liquid and solid states. Complete solubility means homogeneous mixtures of two or more kinds of atoms are formed in the solid state. The more abundant atomic form is referred to as the solvent, and the less abundant atomic form is referred to as the solute. For example, brass is a solid solution of copper (64%) and zinc (36%) so that copper is the solvent and zinc is the solute.

There are two types of solid solutions: substitutional solid solutions and interstitial solid solutions. Substitutional solid solutions, which can be either ordered or disordered, are formed when solvent atoms in the parent metal's crystal structure are replaced by solute atoms. For example, copper atoms may substitute for zinc atoms without disturbing zinc metal's face-centered cubic (FCC) crystal structure. For complete solid solubility, the two elements should have the same type of crystal structure, and for extensive solid solubility, the difference in atomic radii between the two elements should be less than 15% [18]. Solid solubility is favored when the two elements have lesser chemical affinity, since compounds form when chemical affinity is high. Generally, compounds that are separated in the periodic table have higher chemical affinity, so elements that are close together tend to form solid solutions.

In interstitial solid solutions, solute atoms enter holes in the solvent atom crystal structure in interstitial solid solutions. Atoms that have atomic radii less than 1 Å tend to form interstitial solid solutions. Carbon, nitrogen, oxygen, and hydrogen are example interstitial solid solution solutes. Intermetallic compounds are formed when one metal (e.g., magnesium) has chemical properties that are strongly metallic and another metal (e.g., antimony, tin, or bismuth) has chemical properties that are only weakly metallic. Intermetallic compounds have higher melting temperatures than either of their parent metals. The higher melting point indicates a strong chemical bond in the intermetallic compound.

Table 2 lists melting points and carbon to metal ratios (C/M) for several monocarbides that have been investigated for use in space reactors [19]. Solid solution carbides are expected to be able to operate for short periods of time at propellant exit temperatures as high as 3200 K and for many hours at exit temperatures of 2600–3000 K. The life-limiting phenomenon for the solid solutions appears to be vaporization at surface temperatures greater than 2900 K.

The highest melting temperatures for most monocarbides occur at C/M ratios that are less than one, and pseudo-binary and pseudo-ternary carbides have their highest melting temperatures for single-phase solid solutions. The melting point for single-phase solid solution carbides has been shown to be 100–700 K higher than the melting temperature for carbides that have formed a separate carbon phase (e.g., [U, Zr] C_x + C) [20, 21]. **Figure 3** shows the solidus curves for ternary mixed carbides of (U, Zr, Nb) C from [20]. This study showed higher melting temperatures for ternary mixtures than for binary carbides of ZrC or NbC with an equal amount of UC.

Carbon to metal ratios were carefully controlled in the Rover/NERVA Nuclear Furnace (NF-1) tests to prevent formation of second phases that significantly reduced melting temperatures of the carbide fuels. A C/M ratio of 0.88–0.95 was targeted for NF-1, (U, Zr) C fuel elements for a proposed maximum operating temperature of 3200 K [22].

Binary alloy	Melting temperature (K)	Carbon to metal ratios
NbC	3873 ± 25 K	0.79
TaC	4258 K	0.89
UC	2803 K	1.00
ZrC	3813 K	0.87

Table 2.
Melting temperatures and carbon to metal ratios of various monocarbides [19].

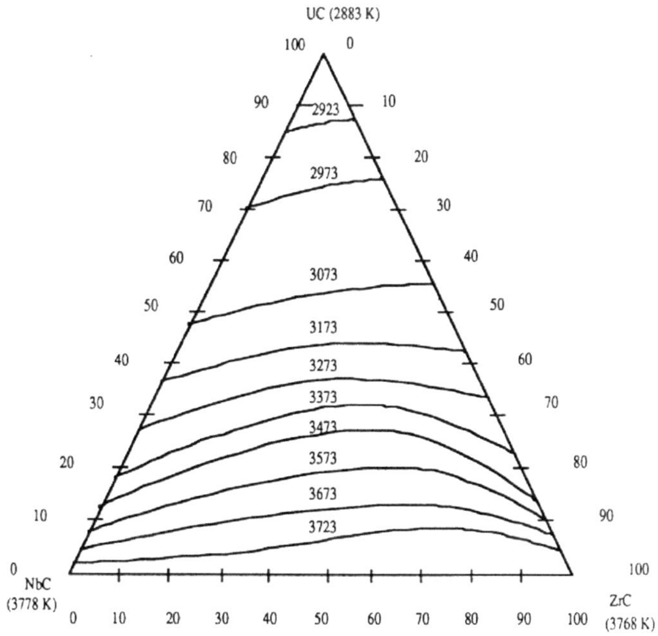

Figure 3.
Solidus curves for ternary mixed carbides of (U, Zr, Nb) C [20].

Fabrication of carbide fuel elements was completed in several steps under Rover/NERVA. First, a mixture of ZrC, UO_2, ZrO_2, graphite flour, and binder was prepared and extruded. Free carbon was removed from the extrusion by leaching with hot flowing hydrogen. The fuel elements were then impregnated with zirconium to varying degrees using a CVD process to produce a single-phase solid solution carbide element that was substoichiometric in carbon. Extrusion of these elements produced severe die wear because of the carbide content, so 19 mm (0.75 in) wide hexagonal elements containing 19 coolant channels could not be directly fabricated through extrusion. Instead, the hexagonal elements were manufactured by first extruding cylindrical fuel forms and machining them to a hexagonal geometry.

All three of the carbide fuel constituents can be mixed, cold pressed, and sintered to fabricate a fuel pin, but mixing all of the components at once makes it difficult to control C/M ratio and prevent the formation of a second carbon phase. Carbide particles also tend to be coarse and require long sintering times at high temperatures in order to produce a homogeneous material.

The major problem with the use of graphite and other carbon-based fuels (e.g., UC, UC_2, [U, Zr] C) in high-temperature space reactor applications is mass loss produced by a number of interrelated and competing physical processes [23]. These processes include the formation of carbon liquids, loss by vaporization, extensive creep, and corrosion during hydrogen exposure. Maximum mass losses typically occurred in moderate-temperature regions of the core (<2000 K). The amount of hydrogen corrosion that occurs is dependent on:

- Reactor operational duration
- Number of fuel duty cycles
- Local material temperatures
- Local hydrogen gas flow conditions
- Fuel location in the reactor
- Reactor power density
- Compatibility of the fuel and coatings

There are four major coupled reactions and/or healing processes associated with hydrogen corrosion:

- Exposure to hydrogen gas
- Nonuniform loading and/or cycling of the fuel
- Radiation exposure
- Creep

The first process is directly associated with chemical corrosion, while the remaining processes affect the amount of cracking that occurs in the fuel, which affects the fuel surface area that is exposed to hydrogen.

Carbon-based fuel materials can experience mass loss by two mechanisms when exposed to hot hydrogen: vaporization (or sublimation) of material constituents at temperatures above carbide vaporization temperatures, and chemical reaction of carbon constituents with hydrogen to form hydrocarbon gas species such as methane (CH_4) and acetylene (C_2H_2). Vaporization occurs at varying rates in the moderate- to high-temperature regions of U-Zr-C fuel, but it is the predominant mass loss mechanism at temperatures greater than 2900 K. Little chemical reaction between hydrogen and carbide materials takes place at temperatures below approximately 1500 K, but chemical reaction losses predominate below the vaporization temperature of carbide materials (i.e., temperatures between 1500 K and 2900 K). The formation of CH_4 becomes increasingly unstable at low to moderate hydrogen pressures and temperatures greater than 1500 K, since C_2H_2 is the more stable compound under these conditions, but the opposite relationship is true for higher pressures [24].

Figure 4 shows the recession rate of $U_{0.1}$ $Zr_{0.9}$ C compared to other compounds as a function of temperature and illustrates the fact that the diffusion rates of carbon and uranium can be substantial at high temperatures. Changes in surface chemical conditions in U-Zr-C materials likely encourage the release of free carbon, since surface composition changes tend to enhance the diffusion of carbon and uranium to the fuel surface because of shifts in the U/Zr/C ratio. These changes were noted during start-up of the Rover/NERVA reactors and were determined to be a predominant contributor to corrosion mass loss. Corrosion mass loss may also degrade fuel surface properties so that particles, such as fuel grains and grain agglomerations, become loosened and erode into the hydrogen gas stream.

The presence of hydrocarbons in the propellant stream tends to decrease the release rate of carbon from downstream fuel surfaces [26]. This effect may be a partial explanation for the lower corrosion rates that were observed in higher-temperature regions of the Rover/NERVA fuel elements. Another partial explanation for the lower corrosion rates may be the healing of surface defects due to material creep at high temperatures. This healing process may reduce hydrogen intrusion into the high temperature fuel regions, but the healing effect may be reduced by radiation damage. Radiation damage can also reduce the thermal conductivity of the fuel, which can produce locally high thermal stresses and

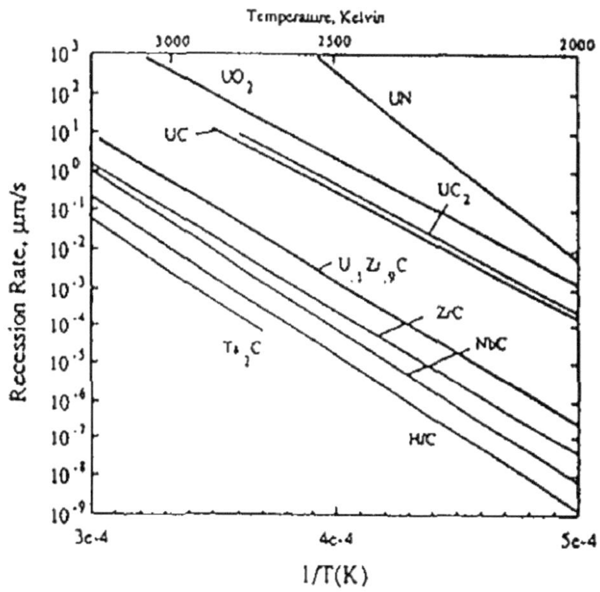

Figure 4.
Recession rate of $U_{0.1}$ $Zr_{0.9}$ C compared to other uranium compounds and refractory carbide materials [25].

corresponding mismatches between stresses in fuel coatings and the fuel substrates. These mismatches in turn encourage the formation of surface coating cracks that enhance hydrogen penetration into the fuel and offset any beneficial effect produced by creep healing.

Another major hydrogen corrosion initiator is nonuniform loading and thermal cycling of the fuel. Nonuniform loadings and material expansion effects were considered to be a major cause of reduced corrosion in higher-temperature fuel regions due to closing of surface cracks [24]. Nonuniform mechanical loading can be produced by:

- The presence of preload stresses during fuel fabrication and application of material coatings. The Rover/NERVA reactors were often designed with tie tubes that had substantial preloads that offset the axial loads produced by high hydrogen pressure differences.

- The presence of nonuniform or unsteady pressures and nonuniform axial temperature distributions. Varying pressure profiles are always present during NTP transient operations, such as during reactor start-up and shutdown.

- Residual stresses produced by fuel cycling. For example, fuel can undergo creep at high temperatures, which may lead to high induced tensile and compressive stresses during fuel cooldown.

It was initially believed during the Rover/NERVA program that radiation effects would be minimal in carbon-based fuels because of carbon's resistance to radiation and the low operation time for NTP systems. This belief turned out to be unfounded because of the high-power densities that are required in NTP reactors. Post-test examinations of the Rover/NERVA reactors showed that radiation damage caused reductions in thermal conductivity and ductility, and these reductions caused cracking that allowed hydrogen to enter the fuel [22].

4. Future work

NASA is once again exploring the feasibility of building and operating nuclear fission systems for use in deep space science and exploration missions. The primary objective for the feasibility studies is to identify systems that can be used to support human missions to Mars, but missions to the outer solar system and beyond are also under consideration. The major barrier to demonstrating a high-performance nuclear propulsion system is developing a fuel that can survive the extreme operating conditions that will be required during space flight missions. The fuel operational characteristics that will need to be satisfied during reactor operations include:

- Minimizing high-temperature hydrogen corrosion

- Minimizing brittle fracture behavior at low temperatures

- Minimizing fuel creep and vaporization at high temperatures

- Minimizing radiation damage that impairs fuel performance

- Managing high transient thermal and mechanical stresses on the fuel during reactor start-up

- Rapid heat transfer from the fuel to the propellant
- Matching coefficients of thermal expansion for different materials used in the fuel to avoid fuel constituent separation during reactor operation
- High uranium loading to allow for use of low-enriched uranium fuel
- Low fuel and reactor system mass to minimize launch costs
- Limiting fuel dissociation and constituent migration during reactor operation
- Limiting cracking of fuel and coatings to minimize hydrogen ingress into the fuel during reactor operations.

As a result of these restrictive requirements, it is likely that whatever fuel is selected will have to operate close to its thermal and mechanical failure limits. There will be little margin for error in system operation, so a significant amount of research and testing will be needed before a safe and reliable system can be built and operated. The majority of future work associated with developing space reactor propulsion and power generation reactors will be associated with designing, building, and operating the equipment and experiments that will build on past testing programs and lead to fuel and reactor qualification and public acceptance.

Author details

Douglas Burns* and Stephen Johnson
Idaho National Laboratory, Idaho Falls, Idaho, USA

*Address all correspondence to: douglas.burns@inl.gov

References

[1] Human Exploration of Mars Design Reference Architecture, SP-2009-566, NASA. 2009

[2] Watson CW. Nuclear Rockets: High-Performance Propulsion for Mars, LANL Report LA-12784-MS. 1994

[3] American Institute of Aeronautics and Astronautics. Nuclear Space Power and Propulsion Systems. Vol. 225. Claudio Bruno Ed.: AIAA; 2008

[4] Schnitzler BG, Borowski SK. Small fast spectrum reactor designs suitable for direct nuclear thermal propulsion, AIAA 2012-3958. In: 48th AIAA/ASME/SAE/ASEE Joint Propulsion Conference and Exhibit, 30 July–01 August 2012; Atlanta, GA. 2012

[5] Koenig DR. Experience Gained from the Space Nuclear Rocket Program (Rover). Report LA-10062-H. Los Alamos, NM: Los Alamos National Laboratory; 1986

[6] DOE/NV. Nevada National Security Site History: Nuclear Rocket Development Station, DOE/NV-707 rev 2. 2013

[7] Bhattacharyya. An Assessment of Fuels for Nuclear Thermal Propulsion. ANL/TD/TM01-22. 2001

[8] General Electric. 710 High-Temperature Gas Reactor Program Summary Report: Vol. I, Report GEMP-600-V1; Cincinnati, OH. 1968

[9] Argonne National Laboratory, Nuclear Rocket Program Terminal Report, Report ANL-7236; Argonne, IL. 1968

[10] General Electric. 710 High-Temperature Gas Reactor Program Summary Report: Vol. III, Fuel Element Development, GEMP-600-V3. 1968

[11] Nieh TG. Solid-state diffusion bonding of tungsten-25 rhenium alloy. Journal of Material Science. 1986;**2**: 2327-2334

[12] Haslett RA. Space Nuclear Thermal Propulsion Program Final Report, Grumman Aerospace Corporation, PL-TR-95-1064. 1995

[13] Butt DP, Wallace TC. The U-Zr-C ternary phase diagram above 2473 K. Journal of the American Ceramic Society. 1993;**76**(6):1409-1419

[14] Dyakov E, Tishchenko M. Manufacture and Tests of the Fuel Elements in Hydrogen, INSPI Contract Report. 1994

[15] Knight T, Anghaie S. Ternary carbide uranium fuels for advanced reactor design applications. In: 7th International Conference on Nuclear Engineering; Tokyo, Japan; April 19-23, 1999 ICONE-7829. 1999

[16] Nightingale RE. Graphite: Advantages, limitations, and applications. Nuclear Science Abstracts. 1966;**20**(21)

[17] Eatherly WP, Piper EL. In: Nightingale RE, editor. Nuclear Graphite. New York: Academic Press; 1962. pp. 21-51

[18] Reddy LK. Principles of Engineering Metallurgy: Solid Solutions. New Delhi, India: New Age International Limited Publishers; 1996

[19] Massalski TB. Binary Alloy Phase Diagrams. Vol. I. Metals Park, Ohio: American Society for Metals; 1986

[20] Tosdale JP. Refractory metal-carbide systems [MS thesis]. Ames, IA: Ames Laboratory, Iowa State University; 1967

[21] Czechowicz DG, Hampel FG, Storms EK. Proceedings of M. S. El-Genk, 8th Symposium On Space Nuclear Power and Propulsion, AIP Conference Proceedings. Vol. 910116. New York, NY: American Institute of Physics; 1991. pp. 1059-1063

[22] Lyon LL. Performance of (U,Zr)C-Graphite (Composite) and of (U,Zr)C (Carbide) Fuel Elements in the Nuclear Furnace I Test Reactor, Los Alamos Scientific Laboratory Report No. LA5398-MS. 1973

[23] Pelaccio DG, Genk MS, Butt DP. Hydrogen corrosion considerations of carbide fuels for nuclear thermal propulsion applications. Journal of Propulsion and Power. 1995;**11**(6):1338-1348

[24] Connell LW. Hydrogen Corrosion of Graphite Nozzles for a Particle Bed Nuclear Rocket, Sandia National Laboratory, SAND 89-0004; Albuquerque, NM. 1989

[25] Matthews RB, Baars RE, Blair HT, Butt DP, Mason RE, Stark WA, et al. Fuels for space nuclear power and propulsion: 1983-1993. In: El-Genk MS editor. A Critical Review of Space Nuclear Power and Propulsion 1984-1993. New York: American Institute of Physics; 1994. pp. 179-220

[26] Taub JM. A Review of Fuel Element Development for Nuclear Rocket Engines, Los Alamos Scientific Laboratory Report LA-5931; Los Alamos, NM. 1975

Section 6

Nuclear Materials and Data and Reactor Core Calculations

Chapter 7

Parallel Algorithm Analysis in Reactor Core Calculation

Pingzhou Ming

Abstract

The reactor core system consists of many materials, involving multi-physics processes, and can be analyzed and simulated at multi-scales. With the evolution of cluster computer, traditional programs and models could be translated into new program structure and modified in detail, so that more complex problems can be solved. Based on existing theory, programs of sub-channels analysis, two-dimensional (2D) method of characteristic (MOC), fuel temperature approximation, and three-dimensional (3D) discrete ordinate method (SN) are developed and analyzed. The different approach is that the reusable software structure of core calculation is established, with more well-defined storage of nuclear data, control layers, and more effective parallel algorithm for computation. The features of parallel algorithm for these programs are listed succinctly in the discussion. Additionally, the corresponding testing on parallel algorithm and computing results are given.

Keywords: reactor core, modeling and simulation, nuclear data, software structure, parallelism

1. Introduction

There are various heterogeneous phenomena in reactor core, which are related to multi-physics and multi-scales modeling and simulation [1]. The core program or software is the important approximation for the reactor core; therefore, mathematics models, numerical algorithms, preprocessing, post-processing, and visualization are sometimes managed in the similar software structure [2]. Based on the existing conditions of calculation, high-performance-computing technology is an important mean to accomplish the simulation of core programs by the cluster computer. For instance, there are several supercomputers (cluster) in ANL and ORNL that provide high-capacity resources for reactor engineering calculation and nuclear-related simulation [3]. Once the numerical algorithm is fixed, the time-to-solution becomes the necessary consideration during the research, and the approach pattern can be summed up in three steps [4] in this field:

1. Develop the single discipline program or multidisciplinary coupling code for the reactor core, which enable higher accuracy.

2. Propose and implement parallel algorithm for the program.

3. Use parallel hardware to run and carry out numerical experiments within acceptable requirements, and then match the reference solution or experimental data.

According to the modeling and simulation for reactor and the features of multi-physics and multi-scale models, the reactor calculation is classified into core calculation and out-core calculation [5]. The core calculation covers considerable subjects, for instance, reactor physics, thermal-hydraulics, dynamics, and fuel performance, for example. For the purpose of research and engineering, different discrete systems and solvers are implemented so as to reveal different phenomena in the core [6–8]. As an example, the DENOVO transport software of ORNL utilizes 160,000 processors to run the benchmark core example in Jaguar supercomputer; up to now the computer has been upgraded to TITAN [9].

On the one hand, the simulation could be accelerated through effective parallelization in different models [10]. On the other hand, the parallel computing makes researchers focus on more research fields except mathematics and physics [11], which constructs one view from nuclear data to top-level application. With the attempt of the M&S content, the core software could be ascribed as the expression and translation of the set of state variables and data structure [12]. Thus, the conjunction of the core problem and parallel computing is from the new requirements of application; furthermore it is driven by the underlying hardware innovation. The Berkeley Parallel Computing Laboratory reviews parallel computing and points out that the parallel elements of different algorithms are the methodology of application driven and reuse. Since various features exist in different layers of software, the parallel algorithm should be designed specifically so as to adapt the parallel hardware [13]. The chapter describes and explains the software structure, underlying nuclear data, and parallel algorithm from a systematic view in the domain of reactor core calculation. Section 2 abstracts the software structure for core calculation. Then, Section 3 explains the basic elements of nuclear data and its association according to the software structure. After that, Section 4 lists the concept and steps of parallel algorithm analysis for core problem in practice. Sample programs are analyzed and discussed concisely in Section 4, and conclusions are given in Section 5.

2. Software structure

As the complex system, the reactor core could be understood by various parts, such as concept models, numerical methods, model uncertainty, and model reuse. Algorithms and software are the center of digital simulation so that the calculation rule is regulated by the program technologies. Sometimes the fixed software structure could be abstracted, and common algorithms could be extracted to be close to the trend of reactor core simulation, which forms one united manner (also known as framework method [14]).

According to the description [12], one general software structure could be divided into seven layers, which covers all aspects from the application to the computer operating system by **Table 1**.

Then, the numerical calculation system is abstracted into multiple layers, which are implemented through controls, modularity, and dynamic interaction. So participants from different professional backgrounds pay more attention to their own business [15]. The core calculation learns from the above contents, and the software structure could be simplified in **Table 2**.

It is effective to develop numerical software based on reusable components or framework, so the discrete model could be modeled through fixed process. On the one hand, the usage leads to reuse. On the other hand, new programming pattern could be merged. For instance, with the help of the underlying framework, MC neutron transport combines the discrete model and algorithm analysis process and

Abstraction layer	Function
Model layer	The model code
Model framework layer	Collection of models for a single application area
Simulator layer	Provides the paradigm for simulation and synchronization usage
Component federation layer	Provides interface code to create one model with multiple sub-models
Load management layer	Within one parallel model execution, collects run information
Ensemble layer	Runs instances in a job and handles scheduling
Operating system/job scheduler layer	Runs independent jobs in parallel

Table 1.
General software structure.

Abstraction layer	Function	Examples
Numerical framework	Provides basic arithmetic function and express data	PETSc, Trilinos, HDF5 NetCDF, BLAS, MPI
Discrete model	The single component with the properties of time, spacing, parallelism, and synchronization	Structured grid Unstructured grid Discrete timeline
Component combination	Provides interface code to create model of multiple components	Geometry control Cross-sectional database Neutron transport solver
Algorithm analysis	Schedule the model in parallel, and process the run data and obtain running information	TAU VTK

Table 2.
Software structure for core calculation.

then uses numerical components and structure to accomplish the transport task easily in the reactor core [16]. Since this integrated software structure method restricts the programming manner of the core calculation and provides reuse mechanism, then in practical, the prototype software NAC4R is designed and constructed that contains basic data operation, basic algebra operation and linear system solver, and so on. As an example of the above data, NAC4R can be used for the parallel analysis process of sample programs that explains the effectiveness and efficiency of the corresponding software structure.

3. Nuclear data process

The underlying nuclear data is the basis of core calculation and also the fundamental elements in the proposed software structure. The reactor physics data is divided into four categories by A. Trkov, namely, (1) basic nuclear data, (2) evaluated nuclear data files, (3) processed nuclear data, and (4) averaged nuclear data [17]. They are the similar data in the thermal-hydraulics analysis, which correspond to the physical property information of the fluid, such as the light water property data in the core calculation [18]. These different types of data involve representation, storage, and computer algorithms. For example, the cross-sectional library of deterministic reactor physics calculation needs group-wise data, such as few-group homogenized cross section in diffusion calculation, with the affection of multiple state variables that include temperature, density, and so on [19]. The class of used nuclear data can be directly computed in the core calculation or be generated in advance. The pre-generated nuclear data (database) is commonly used in a two-step method of core calculation, one is multidimensional interpolation table, the other is parameterized library, and they have different effects on computation and parallel algorithm [20].

3.1 Multidimensional interpolation table

The table is the indirect format of cross-sectional parameters, with the grids inside the value range for each state variable. The core calculation uses interpolation algorithm to obtain the nuclear data.

Assume there is only one state variable ρ (such as coolant density), and its value range is $[a, b]$, then the grid of a single state variable is divided as in Eq. (1):

$$\{\rho_i\}_{i=1}^{N+1}, \rho_i \in [a, b] \tag{1}$$

where ρ_i is not necessarily uniform distribution. The corresponding parameters at each discrete grid can be obtained by the lattice program as follows in Eq. (2):

$$\left\{\sum_i\right\}_{i=1}^{N+1}, \sum_i := \sum(\rho_i) \tag{2}$$

In this way, the continuous parameter values can be calculated by polynomial interpolation method in the range of state variable. The coefficients of interpolation polynomials are stored in the table:

$$\tilde{\sum}(\rho) = P(\rho) = \sum_{i=0}^{N+1} a_i \rho^{i-1} \tag{3}$$

3.2 Parameterized library

The library method is the complex function model of nuclear data that is related to various state variables. To build an accurate model, many factors need to be considered, and the range of each state variable is also defined separately:

$$\sigma = f(\rho, T, P, \ldots) \tag{4}$$

Generally, the tabulation method uses space for time, and the method is relatively simple, but the storage mode of nuclear data has a greater impact on the parallel algorithm of the core calculation. The method consumes more computing time, but the data storage is smaller, and it is easy to improve and modify the model so that there is less impact on the parallel algorithm.

The software implementation of the nuclear data corresponds to the model layer and model framework layer mentioned in Section 2. **Figure 1** shows two abstract forms of the nuclear data. The left side that represents the processing of each kind of nuclear data is independent, and the right side explains the design of consistent data attributes, and formats are independent underneath and dependent on top in the fixed software structure so that each type of nuclear data only needs format conversion and minor programming. This understanding of the underlying nuclear database focuses on software reuse and performance, which can fully reuse various

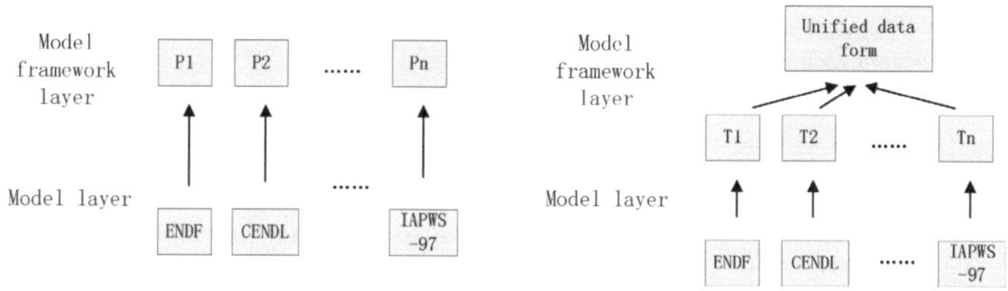

Figure 1.
The association between underlying nuclear data and software structure.

known nuclear data that have fine readability and correctness. This association also enables independent design and programming of core calculation at all levels, which makes the following parallel algorithm analysis match it.

4. Parallel algorithm analysis

Amdahl law stipulates the speedup ratio limit for parallelism, and Gustafson law quantifies the acceleration effect [21]. To be equivalent to Amdahl and Gustafson, measurability of calculation models the performance. Assume that the running time of one computing problem is $T(N,x)$, which represents the running time for the scale of problem x. Then the running efficiency could be utilized to prove the effectiveness of parallel core program from two categories.

4.1 Strong measurability

The change of computing time varies with the increase of processor cores, also is known as speedup ratio:

$$S_{strong} = \frac{T(1,x)}{T(N,x)} \tag{5}$$

4.2 Weak measurability

The problem size of each processor or process is fixed, and then the statistic of the running time varies as the processors change:

$$S_{weak} = \frac{T(1,x)}{T(N,x \cdot N)} \tag{6}$$

The deterministic calculation of the reactor core mainly makes the $S_{strong,N}$ as the variation of processor cores. In practical deterministic core, computation has features that the problem scale is related to the physical form. For instance, when the characteristic rays are fixed on every processor, the MOC transport would increase new characteristic rays and achieve new arrangement if the processors are added. The data expression and computational content may change with different program patterns. On the contrary, nondeterministic calculation of reactor core, such as the MC method, the computational features would remain unchanged so that the problem takes more emphasis on the statistics of weak measurability.

The steps proposed by Ian Foster is the most typical method for the parallel algorithm design and programming [22] (also known as PCAM method).

	Designed and implemented algorithm	Algorithm is not designed and implemented
Working step	S1: executes and obtains the statistical parameters of parallel performance on a fixed running platform S2: judges whether the requirements are met. If not, go to step 3. Otherwise, ends the analysis S3: designs or improves the existing parallel algorithm and then goes to step 1	S1: executes and obtains the statistical parameters of parallel performance on a fixed running platform S2: identifies the parallel features of the algorithm S3: determines whether the features have reusable patterns in the software structure (here refers to NAC4R). If so, reuses them directly, turns to step 5, if not, turns to step 4 S4: the new parallel algorithm is designed and programmed S5: runs on the fixed platform and obtains the statistical parameters S6: judges whether the requirements are met. If not, goes to step 3. Otherwise, ends the analysis and merges the new parallel algorithm into NAC4R as a reusable pattern
Key issues	1. The performance is closely related to the problem 2. The features depend on the abstraction of actual core numerical calculation that is often distinguished by the knowledge and experience of developers	

Table 3.
Two different analysis steps.

1. Partition identifies computing content that can be executed in parallel.

2. Communication determines the communication data of each parallel task.

3. Aggregation merges and adjusts parallel tasks and communication.

4. Mapping assigns the aggregated task to entities on which actual parallel programming techniques depend (process or thread).

At present, the cluster computers are often in heterogeneous development environment, and parallel algorithms are divided into many kinds such as pipeline parallelism, mater-slave parallelism, and so on. To facilitate engineering specifications, the usually analysis steps of parallel algorithm are divided into two cases in **Table 3**.

It can be concluded that the parallel analysis includes interpretation, design, and performance verification. The parallel analysis in reactor core calculation summarized here emphasizes to meet the application requirements, and then the common features of the algorithm are usually abstracted and recorded concisely, so that the analysis can be understood clearly and simply.

5. Samples research

Interconnections exist in the multi-physics process and software structure in reactor core. For solving according to physics-mathematics model, various core software has been developed. There are two main ideas for developing these procedures. One is to make full use of the existing programs in the world. The other is to redesign and program the program structure and function. The key parallel algorithms of single discipline sample program CORTH, KYLIN2, K-MOD, and Hydra are identified and programmed, which come from the typical examples of the reactor core system, computing work through the research career of the author.

5.1 Sub-channels analysis

The CORTH program utilizes the uniform flow model of four equations to process two-phase flow in the reactor core. The energy conservative equation could be abbreviated as the manner of structured grids, which is easy to use in finite difference and form the linear system. For the description, CORTH could be divided as components in **Table 4**.

The features of the parallel algorithm are listed summarily to make the program refinement practical.

1. The reactor core is divided along the radial direction so that each process stores the data of some sub-channels. The overlap method could also be used to reduce data communication.

2. The solution of linear system solving can be quick in the model. The coefficient matrix of the conservative system is symmetric, and it can be constructed simultaneously together with the right-hand vector of the system.

3. The PETSc and Aztec parallel solver can be ported and optimized inside NAC4R so that solvers could be used and hybrid programmed directly and easily.

The bottleneck of the parallel algorithm design is related to the components of the coefficient update, linear system solve, and mesh matching. Parallel scheme was realized in NAC4R with the features mentioned above. The ACP1000 model (177 fuel assemblies, there are 264 fuels and 25 guide tubes in each assembly, axial segments are 54, and there are 50868 sub-channels in radial direction) is used for performance validation. Cluster processors are chosen as 4, 6, 18, 27, 36, 54, and 108, which is the divisor of 50868 as the Aztec solver needs that the number of processes matches the tasks. The expected performance of strong measurability is improved by more than five times.

Performance conclusion: The trend of processor change shows that using about 30 processors, it has met the expectation in **Figure 2**, with the speedup ration about eight times. Not only the conservation system can be constructed in parallel, but the Aztec solver can also solve the matrix with the size of 50868 efficiently.

5.2 2D MOC

The KYLIN2 program carries out the 2D neutron transport calculation by MOC over the assembly region. In the condition of steady situation, multigroup neutron

Components	Model equation	Function
Preprocessing		Reads external input and translates into data structure in memory
Pre-computation		Calculates some parameters and provides initial values for iteration
Iterative control		Controls and manages cyclic calculation
Coefficient update	$f(G, H, \dots)$	Updates parameters of thermal-hydraulics
Linear system solve	$x = A^{-1}b$	Defines linear system and its computation
Mesh matching	$m_{i,j,k} = m_{i1,j1,k1} \cup \dots \cup m_{iN,jN,kN}$	Maps the sub-channel, meshes, and processes

Table 4.
Abstract components in CORTH.

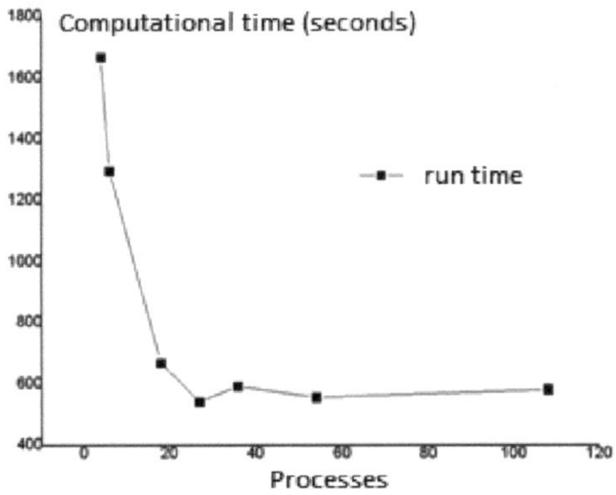

Figure 2.
Time statistics of model ACP1000 for CORTH.

transport equation could be expressed by the ODE along the characteristic ray. We can obtain the direct analytical solution of the ODE as in Eq. (7):

$$\psi_{out,m,i,k} = \psi_{in,m,i,k} e^{-\sum_i S_{m,i,k}} + \frac{Q_i}{\sum_i}\left(1 - e^{-\sum_i S_{m,i,k}}\right) \tag{7}$$

where $\sum_i$ and Q_i denote the macro-transport cross section and the neutron source, respectively. The length $S_{m,i,k}$ is the segment that is truncated with the position m-th azimuth angle, k-th characteristic ray, and the i-th mesh. When source iterative method is used, the iterative procedure will repeat the above calculation process after assigning initial guess values to the flux. As shown in **Table 5**, sample program KYLIN2 could be abstracted into different parts.

Sweep computation follows the loop structure, with each characteristic ray, neutron groups, azimuth angles, and polar angles. Moreover, the intermediate solution of KYLIN2 is related to the angular direction so that the structure can be described as follows.

```
set current←0
set mesh_angular_flux←0
for (g, ray, p) in (Groups, Rays, Polars):
    current ← computeForwardCurrent(g, ray, p)
    mesh_angular_flux ← sweepForward(g, ray, p, current)
    current ← computeBackwardCurrent(g, ray, p)
    mesh_angular_flux ← sweepBackward(g, ray, p, current)
end for
```

Components	**Model equation**	**Function**
Geometry description		Discrete result of 2D geometry
Nuclear data	$(\sum_i)$	Stores cross-sectional data with meshes
Characteristic rays	$(S_{m,i,k})$	Generates and manages rays
Sweep computation	$\overline{\psi}_{m,i,k} \leftarrow f(\alpha, e^{\beta})$	Traverses rays and solves angular neutron flux
Source update	$Q_i \leftarrow f(\sum_i, \overline{\psi}_{m,i,k}, \omega)$	Updates the transport source
Task schedule		Controls the computational flow, such as BSP parallel model

Table 5.
Abstract components in KYLIN2.

The above components are redeveloped in NAC4R, and the features of the algorithm are listed here:

1. Iterate that each ray could be in parallel if some critical operations are added, so both MPI and OpenMP can be used for the sweep process.

2. If rays are non-modular arranged, the differences of working load of each ray are not obvious. Since the assembly problem sometimes uses the nuclear data with some energy groups, parallelism on neutron groups could be a better choice, and OpenMP can avoid data communication during the sweep procedure.

3. The callings of exponential function will affect the efficiency.

4. In order to adapt coarse mesh acceleration, reserved interfaces should be kept inside NAC4R so that parallel linear system solvers are used.

Here OpenMP guide sentences are utilized to implement parallelism based on loop structure. The critical operation exists in the cumulative sum of forward and backward sweep when each ray is accessed. C5G7-MOX assembly and ACP1000-UO2 assembly examples are used to test, and more than five times better performance are expected. The statistical calculation time is shown in **Table 6**.

Performance conclusion: OpenMP parallelism (multi-threads) can achieve the rational accelerated effect and meet the expectation. There are more speedup ratios for the ACP1000-UO2 as the neutron group parallelism strategy is used, and some fluctuation exists in the performance gain of shared memory parallelism since there are some management overheads by multi-threads.

5.3 Fuel temperature approximation

The calculation of the fuel temperature usually involves the temperature of the outer surface and the center, and then the effective temperature of the fuel could be obtained for rod-like geometry. The geometry is divided into different sections to solve the heat conduction equation in radial direction, with different approximations are taken account. The fuel rod of PWR is usually delimited as three sections in some programs, and the 1D conduction equation is constructed at each area. After knowing the boundary temperature, the Rowland's equation could be used to compute the effective temperature.

Fuel temperature approximation is necessary for the reactor core coupling calculation, but reasonable boundaries come from other discipline procedure. K-MOD coupling program has been realized, and according to the above description, fuel temperature approximation could be abstracted in **Table 7**.

Thermal feedback controller carries out the function of management and as an external module that will be called repeatedly in K-MOD. The features of the parallel algorithm are less obvious and can be listed in two points: (1) the computing

Example	One core	Four cores	Eight cores	Speedup ratio
C5G7-MOX	1655.28 s	562.20 s	295.01 s	5.61
ACP1000-UO2	19515.38 s	5091.92 s	2581.34 s	7.56

Table 6.
Parameters and time results of two assembly examples.

Components	Model equation	Function
Thermal feedback controller	$f(T_{coolant}, T_{fuel}, \phi)$	Main control component that processes computation in each discrete time step
Fuel temperature calculation	$T_{fuel} \leftarrow f(T_{coolant}, P, G, t)$	Solves the conduction equation by iterative method and obtains the temperature
Cross-sectional update interface	$\Sigma \leftarrow f(T_{fuel}, T_{coolant}, \phi)$	Data exchange interface for the temperature and cross section

Table 7.
Abstract components in K-MOD.

granularity is small and the single cyclic time is short and (2) the computing pattern could be abstracted and reused if lower level parallel method is researched; thus, SIMD technology is suitable for programming.

In the sample program K-MOD, fuel temperature calculation could be abstracted into three classes, such as op_basic represents the basic arithmetic, but op_math stands for the compound operation, such as square, logarithm, and so forth.

vector<int> a; vector<int> b; $c \leftarrow op(a, b)$	vector<float> a; vector<float> b; $c \leftarrow op\_basic(a, b)$	vector<float> a; vector<float> b; $c \leftarrow op\_math(a, b)$
compile option	icpc -O0 -march = native -mno-sse	
compile option with SIMD	icpc -O0 -xsse2	

The former deals with integer data, while the latter two deal with floating-point operations. The problem of rod inserting and lifting in real PWR core is tested and compared. The similar imitated structure solves 50,000 times the same as the K-MOD program, with the SIMD hand optimization, which is based on data structure named __m128i and __m128. The time results are shown in **Table 8**.

Performance conclusion: The statistical results of multiple runs show that SIMD scheme has performance bonus for the third class of calculation content; however, other classes get no gain. Looking up Intel-related documents, it is found that SIMD programming depends on the specific hardware and instruction set usage and depends on the compiler behavior, which is not suitable for manual SIMD rewriting with care in practice.

5.4 3D SN

The sweep operation of discrete ordinate method iterates every angle direction in each octant, which has famous KBA parallel sweep algorithm for structured grids. 3D SN method translates Boltzmann transport equation into the matrix system of flux moment as in Eqs. (8) and (9) in the sample program Hydra:

	First class	Second class	Third class
Serial	93 ms	89 ms	367 ms
SIMD parallelism	112 ms	135 ms	152 ms

Table 8.
Time results of three classes on parallel contents.

Components	Model equation	Function
Geometry storage		Manages structured meshes in XYZ coordinate system
Nuclear data	$(\sum_i)$	Records cross-sectional information
Operator process	(H, Γ)	Constructs mapping between the operator and other parameters
Sweep computation	$\phi \leftarrow H^{-1}Q$	Matrix operation by loop structures
Source update	$Q, S \leftarrow \Gamma\phi$	Updates the transport source

Table 9.
Abstract components in Hydra.

$$L\psi = MS\phi + \overline{Q} \tag{8}$$

$$\phi = D\psi \tag{9}$$

where $\overline{Q}$ and L denote the source term and the difference operator, respectively. Discrete operator of flux moment matrix M need to multiply the cross-sectional data, such as the scattering matrix S. The total meshes are (I, J, K) in each angle direction, and then every section (I_a, J_b, K) that are divided could be mapped into corresponding processes in parallel. According to the above description, SN transport calculation could be abstracted as in **Table 9**.

There are multiple sweeping levels in the SN method, and the section (I_a, J_b, K) could be represented by the loop structure in every octant, and the sweeping procedure can be described as follows:

I,J,K←0,1,2
for k in K:
$\phi_{up,j} \leftarrow \phi_{in}$
for (i, j) in (I_a, J_b):
$\phi_{psi} \leftarrow \phi_{up}$
call $T\left(i, j, k, \phi_{psi}\right)$
$\phi_{up} \leftarrow \phi_{psi}$
end for
$\phi_{out} \leftarrow \phi_{up,j}$
end for

Since there is a lack of parallel programming for this loop structure in Hydra, the features of the algorithm are listed.

1. The extrapolation model, such as the diamond weight difference model, has an independent subroutine T, which is only related to mesh (i, j, k). Flux moment variables ϕ_{up} and ϕ_{psi} depend on the calculation order, so the sweep is one pipeline structure on the (I_a, J_b).

2. When the scale of loop structure is small, shared memory parallelism is fit for pipeline structure to improve efficiency, which has theoretical gain $\frac{ij}{i+j-1}$.

According to the identified features, OpenMP guide sentences are utilized, and lock array done [I][J] is used to ensure data consistency.

```
#pragma omp parallel for
for j in d_begin[J]:d_end[J]
  for i in d_begin[I]:d_end[I]
    wait(done[j][i])
φ_psi ← φ_up
    call T(i,j,k,φ_psi)
    done[next(j)][i] = 1
  end for
end for
```

Benchmark	NAC4R (one thread)	NAC4R (eight threads)
KUCA reactor	3421.7 s	1935.8 s
IAEA pool reactor	5824.9 s	3551.3 s

Table 10.
Time results by the compile level -Oo.

Table 10 collects the running time of two benchmark example, which utilizes the optimization level -O0, and only one MPI process is fixed during the experiment.

Performance conclusion: The speedup ratio is almost 1.8 and 1.6, which illuminates that the pipeline structure can gain performance bonus with small scale (I_a, J_b) for the SN method, such as the sample program Hydra. It can be further expanded to the parallel strategy of MPI + OpenMP.

6. Conclusion

Digital computers are used to simulate and analyze reactor core. For introducing parallel computing and promoting the efficiency, the core calculation with the integrated software structure is explored, which can provide the methodology of nuclear data, control layers, and parallel algorithms. The main software structure is discussed, and multiple parallel algorithms are analyzed concisely for sample core programs, with the corresponding conclusions as follows:

1. Integrated software structure (framework method) is benefited for the development of core software, and the similarity achieves the goal of reuse. For example, the underlying nuclear data could be managed as the effective manner and mapped to application. With the development of prototype software NAC4R, the research could be carried out in different control layers and demonstrate validity.

2. Parallel algorithms are refined as well as identify the parallel features that exist in the core model based on the existing theory. The features of four sample programs are described succinctly, which work well with the software structure. Parallel programming enables the expected efficiency improvement if some rules are obeyed, and the relevant performance results are given summarily.

It is necessary to establish systematic knowledge and suggestions for software approach and parallel algorithm development in reactor core calculation. On the one hand, it can integrate reusable patterns into computation, such as the instance of prototype software NAC4R. On the other hand, it can explore and record more specific parallel algorithms in this field.

Acknowledgements

The author would like to thank relevant scholars for their suggestions in NPIC, and also thank for the sample programs and computing resources provided by the 9th professional group in the design institute of NPIC. The research is supported by the discovery fund of NPIC (Number: 2019-06).

Nomenclature

A	coefficient matrix
x	length or position, m
k	effective multiplication factor
t	time, s
L	the matrix of Laplace operator
M	the matrix of flux moment
N	number of processors
Q	neutron source, $1/m^3s$
s	segment length of the characteristic ray, m
S	speedup ratio

Greek letters

Δ	difference
ρ	density, kg/m^3
ϕ	neutron flux density, $1/m^2s$
ψ	angular neutron flux density, $1/m^2s$
$\sum$	neutron macroscopic cross section, 1/m
ω	weight factor
ε	error threshold
Γ	operator

Subscripts/superscripts

s	serial
g	neutron energy group
p	parallel
i, j, k, z	mesh position
m	azimuth angle
v	volume
f	fission cross section
psi	temporary data
up	upstream data

Abbreviations

2D	two dimensional
3D	three dimensional
ACP1000	Advanced China Power 1000 MW

BLAS	Basic Linear Algebra Subprograms
CASL	Consortium for Advanced Simulation of Light water reactors
CORTH	CORE thermal-hydraulics
DENOVO	new three-dimensional parallel discrete ordinates code in SCALE
HDF	Hierarchical Data Format
IAEA	International Atomic Energy Agency
KBA	Koch-Baker-Alcouffe parallel algorithm
KYLIN2	advanced neutronics lattice code of NPIC
K-MOD	Kinetic Method of Demo
MOC	method of characteristics
MPI	message passing interface
NAC4R	New Architecture Computing for Reactor Core
NetCDF	scientific data NETwork Common Data Format
OpenMP	Open Multi-Processing
PETSc	Extensible Toolkit for Scientific Computation
PWR	Pressurized Water Reactor
SN	Discrete Ordinate method
SIMD	Single Instruction Multiple Data
TAU	Tuning and Analysis Utilities
VTK	Visualization ToolKit

Author details

Pingzhou Ming
Nuclear Power Institute of China, Chengdu, China

*Address all correspondence to: mingpz@mail.ustc.edu.cn

References

[1] Mylonakis AG, Varvayanni M, Catsaros N, et al. Multi-physics and multi-scale methods used in nuclear reactor analysis. Annals of Nuclear Energy. 2014;**72**:104-119. DOI: 10.1016/j.anucene.2014.05.002

[2] Arkadov GV, Zhukavin AP, Kroshilin AE, et al. The virtual digital nuclear power plant: A modern tool for supporting the lifecycle of VVER-based nuclear power units. Thermal Engineering. 2014;**61**(10):697-705. DOI: 10.1134/S0040601514100012

[3] Baker C, Davidson G, Evans TM, et al. High Performance Radiation Transport Simulations: Preparing for TITAN. SC'12: Proceedings of the International Conference on High Performance Computing, Networking, Storage and Analysis. Salt Lake City, UT, USA: IEEE; 2012. DOI: 10.1109/SC.2012.64

[4] Weber DP, Sofu T, Yang WS, et al. High-fidelity light water reactor analysis with the numerical nuclear reactor. Nuclear Science and Engineering. 2007; **155**(3):395-408. DOI: 10.13182/NSE07-A2672

[5] Okumura K, Oka Y, Ishiwatari Y. Nuclear reactor calculations. In: Nuclear Reactor Design. Japan: Springer; 2014. pp. 49-126. DOI: 10.1007/978-4-431-54898-0_2

[6] Larsen EW. An overview of neutron transport problems and simulation techniques. In: Computational Methods in Transport. Berlin/Heidelberg: Springer; 2006. pp. 513-534. DOI: 10.1007/3-540-28125-8_26

[7] Saha P, Aksan N, Andersen J, et al. Issues and future direction of thermal-hydraulics research and development in nuclear power reactors. Nuclear Engineering and Design. 2013;**264**:3-23. DOI: 10.1016/j.nucengdes.2012.07.023

[8] Edsinger K, Stanek CR, Wirth BD. Light water reactor fuel performance: Current status, challenges, and future high fidelity modeling. JOM. 2011; **63**(8):49-52. DOI: 10.1007/s11837-011-0138-7

[9] Joubert W, Archibald R, Berrill M, et al. Accelerated application development: The ORNL Titan experience. Computers and Electrical Engineering. 2015;**46**:123-138. DOI: 10.1016/j.compeleceng.2015.04.008

[10] Gaston DR, Permann CJ, Peterson JW, et al. Physics-based multiscale coupling for full core nuclear reactor simulation. Annals of Nuclear Energy. 2015;**84**:45-54. DOI: 10.1080/00411450.2014.927364

[11] Prabhu P, Kim H, Oh T, et al. A survey of the practice of computational science. In: 2011 International Conference for High Performance Computing, Networking, Storage and Analysis (SC). IEEE; 2011. pp. 1-12. DOI: 10.1145/2063348.2063374

[12] Carothers C, Ferscha A, Fujimoto R, et al. Computational challenges in modeling and simulation. In: Fujimoto R, Bock C, Chen W, Page E, Panchal JH, editors. Research Challenges in Modeling and Simulation for Engineering Complex Systems. Cham: Springer; 2017. pp. 45-74. DOI: 10.1007/978-3-319-58544-4.ch4

[13] Keutzer K, Massingill BL, Mattson TG, et al. A design pattern language for engineering (parallel) software: Merging the PLPP and OPL projects. In: Proceedings of the 2010 Workshop on Parallel Programming Patterns. ACM; 2010. p. 9. DOI: 10.1145/1953611.1953620

[14] Sbalzarini IF. Abstractions and middleware for petascale computing and beyond. International Journal of

Distributed Systems and Technologies. 2010;**1**(2):40-56. DOI: 10.4018/jdst.2010040103

[15] Mo Z, Zhang A, Cao X, et al. JASMIN: A parallel software infrastructure for scientific computing. Frontiers of Computer Science in China, Beijing. 2010;**4**(4):480-488. DOI: 10.1007/s11704-010-0120-5

[16] Bergmann RM, Vujić JL. Algorithmic choices in WARP–A framework for continuous energy Monte Carlo neutron transport in general 3D geometries on GPUs. Annals of Nuclear Energy. 2015;**77**:176-193. DOI: 10.1016/j.anucene.2014.10.039

[17] Trkov A. From basic nuclear data to applications. In: Workshop on Nuclear Data and Nuclear Reactors: Physics, Design and Safety; 2000

[18] Pavel T, Waltar A. Nuclear data and cross section processing. In: Fast Spectrum Reactors. Boston, MA: Springer; 2012. pp. 77-109. DOI: 10.1007/978-1-4419-9572-8.ch5

[19] International Atomic Energy Agency. Thermophysical Properties Database of Materials for Light Water Reactors and Heavy Water Reactors. IAEA TECDOC Series No. 1496(IAEA-TECDOC-1496). Vienna: International Atomic Energy Agency; 2006. Available from: https://www-pub.iaea.org/MTCD/Publications/PDF/te_1496_web.pdf

[20] Sánchez-Cervera S et al. Optimization of multidimensional cross-section tables for few-group core calculations. Annals of Nuclear Energy. 2014;**69**:226-237. DOI: 10.1016/j.anucene.2014.02.013

[21] Hill MD, Marty MR. Amdahl's law in the multicore era. Computer. IEEE; 2008;**41**(7):33-38. DOI: 10.1109/MC.2008.209

[22] Peter P. An Introduction to Parallel Programming. San Francisco: Morgan Kaufmann; 2011. 392 p. DOI: 10.1016/C2009-0-18471-4